Jules Laforgue and Poetic Innovation

For George,
for everything

Jules Laforgue
and Poetic Innovation

ANNE HOLMES

CLARENDON PRESS · OXFORD
1993

Oxford University Press, Walton Street, Oxford OX2 6DP
Oxford New York Toronto
Delhi Bombay Calcutta Madras Karachi
Kuala Lumpur Singapore Hong Kong Tokyo
Nairobi Dar es Salaam Cape Town
Melbourne Auckland Madrid
and associated companies in
Berlin Ibadan

Oxford is a trade mark of Oxford University Press

Published in the United States
by Oxford University Press Inc., New York

British Cataloguing in Publication Data
Data available

Library of Congress Cataloging in Publication Data
Holmes, Anne, 1929-
Jules LaForgue and poetic innovation / Anne Holmes.
1. Laforgue, Jules, 1860-1887—Technique. 2. Experimental poetry—France—History and criticism. 3. Poetics. I. Title.
PQ2323.L8Z67 1993 841'.8—dc20 92-27745
ISBN 0-19-815876-9

Typeset by Downdell, Oxford
Printed in Great Britain
on acid-free paper by
Bookcraft Ltd,
Midsomer Norton, Bath

. . . car la destinée d'un artiste est de s'enthousiasmer et se dégoûter d'idéaux successifs.

Laforgue

Je songe à une poésie qui serait de la psychologie dans une forme de rêve, avec des fleurs, du vent, des senteurs. D'inextricables symphonies avec une phrase (un sujet) mélodique, dont le dessin reparaît de temps en temps.

Laforgue

. . . the truest poetry is the most feigning.

Shakespeare

I suppose I always try to write the truth . . . I think that one of the great criticisms of poets of the past is that they said one thing and did another, a false relation between life and art. I always try to avoid this . . .

Larkin

Preface

MY debts of gratitude belong to two periods—an early and a recent one. In the early group I must mention Claude-Edmonde Magny, who first introduced me to French Symbolism and to Laforgue, and Mme Georges Jean-Aubry, who generously allowed me to work in Paris on the Laforgue manuscripts belonging to her late husband before the collection was, sadly, dispersed. My greatest debt is to Robert Bolgar of King's College, Cambridge, the most acute, most stimulating, and kindest of research supervisors, a rather unacknowledged authority on Laforgue, and a model, no doubt to many besides myself, of true scholarship.

In the second category I am grateful to the Principal and Fellows of Hertford College, which has, for some years now, provided me with a very pleasant academic base; to the staff of the Bibliothèque Littéraire Jacques Doucet in Paris and of the library of the Taylor Institution in Oxford for their unfailing helpfulness; to Virginia Llewellyn-Smith, who encouraged this project; to Andrew Lockett, who has kindly seen the book through the Press; and to Nathalie Roberts for her expert typing. I am also indebted to three anonymous advisers to the Press for their valuable comments. By asking me to contribute to his centenary collection of essays, J. A. Hiddleston spurred me on to write this book, and I am grateful to him also for permission to reproduce in Chapter 5 the substance of that essay.

A century after his death, Laforgue's star now seems to be in the ascendant. I hope this study will contribute to the growing recognition of him as one of the most interesting and important poets who wrote in nineteenth-century France.

Oxford A.H.

1992

Contents

Abbreviations

AH	Jules Laforgue, *Œuvres complètes*, i. *L'Âge d'homme*, eds J.-L. Debauve, D. Grojnowski, P. Pia and P.-O. Walzer (Lausanne, 1986).
FV	Jules Laforgue, *Feuilles volantes*, ed. D. Grojnowski (Paris, 1981).
Hidd. LA	*Laforgue aujourd'hui*, ed. J. A. Hiddleston (Paris, 1988).
LA	Jules Laforgue, *Lettres à un ami* (1880–6), ed. G. Jean-Aubry (Paris, 1941).
ML	Jules Laforgue, *Moralités légendaires*, ed. D. Grojnowski (Paris and Geneva, 1980).
MP	Jules Laforgue, *Mélanges posthumes*, ed. P. Bonnefis (Paris and Geneva, 1979).
NRF	*Nouvelle Revue française*
OC	Jules Laforgue, *Œuvres complètes*, ed. G. Jean-Aubry (Paris, 1922–30).
PC	*Jules Laforgue: Poésies complètes*, ed. P. Pia (Paris, 1970).

Quotations from *Le Sanglot de la terre*, *Les Complaintes* and other early poems are in the form given in *AH*. Quotations from *L'Imitation de Notre-Dame la lune*, *Des Fleurs de bonne volonté*, and *Derniers Vers* follow the texts in *PC*.

The Literary Situation

THE literary scene in the early 1880s, when Laforgue began his career as a writer, looked very different to him from the picture we have of it nowadays. Certain writings that now appear of crucial importance, both in general and in relation to his work, had not been published, and others that had were unknown to him. The image one forms of one's contemporaries is partial, limited, and confused, and this was particularly so in the case of Laforgue, who came to Paris in 1879[1] as a relatively uninitiated outsider who had to discover his mentors for himself.

Rimbaud's *Illuminations*, a work that perhaps more than any other seems to us to mark the watershed between the old and the new, had been written in the early 1870s, but its poems were not published until 1886, a few at a time in the pages of the literary review *La Vogue*, where, late in his career, Laforgue discovered them and was amazed by them. He would not, however, have had the advantage that we have of being able to interpret them with the backing of the *lettres du voyant*, where Rimbaud sets out his poetic aims in phrases which still appear novel and compelling.

Mallarmé was an established figure by 1880, but, even so, Laforgue's knowledge of him was relatively slight. As late as 10 November 1885 he wrote impetuously to him, begging him to publish a volume of his poems: 'Quand pourrons-nous avoir à nous, dûment rassemblé et fixe, le trésor de vos poèmes en vers et en prose, pour les étudier, les goûter par tous les temps, les emporter etc. pour nous faire enfin humainement et d'un jet idée de qui vous êtes?'[2] And, possibly on the same day, he confessed to Kahn: 'En somme qu'est-ce que j'ai lu? "Prose pour des Esseintes", "L'Après-midi d'un faune", *en passant seulement et pas à mon aise*, et jadis, tout jadis les proses de la *Rep. des lettres* et les vers du "Parnasse"!'[3] (my emphasis). The four prose poems by Mallarmé that had been republished in the *République des Lettres* in 1875 were poems of 1864. Paul Bourget introduced him to them in 1880 or 1881, and they clearly made a deep and lasting impression. Of the four ('Plainte d'automne', 'Frisson d'hiver', 'Un spectacle interrompu', and 'Le Phénomène futur'), he preferred 'Plainte d'automne' in 1881, no doubt

[1] He had lived in the outskirts of Paris since 1876, but it was not until about 1879 that his family moved to a fairly central address.

[2] *LA* 128–9. Published in facsimile only.

[3] Ibid. 127.

because he could relate it directly to his personal experience and to his creative ambitions at the time. A comparatively simple poem, it evokes an autumnal mood of melancholy by association with solitude and death. Laforgue had also tried his hand at this theme, and felt that he had failed dismally: 'Je ne suis pas parvenu à faire sentir la mélancolie des après-midi d'automne. Je me rappelle certain poème en prose de Mallarmé, sur l'automne, voilà!'[4] By 1886, when these poems were republished again in *La Vogue*, he had become more sophisticated; he no longer liked the anecdotal and emotional 'Plainte d'automne', and praised instead the more complex 'Frisson d'hiver': 'je n'aime pas le premier poème qui finit par une pointe, mais bien le second où il y a des choses adorables et même défaillantes.'[5] The poem employs a contrapuntal refrain technique very similar, as we shall see, to that used by Laforgue in many of his *Complaintes*. He also, appropriately, found 'Le Phénomène futur' 'très ingénieux'[6] and, presumably, somewhat obscure.

Early in 1882 Laforgue made the cryptic remark (cryptic since the two poets are in many ways dissimilar in both form and subject-matter): 'Je deviens (comme forme) . . . mallarméen',[7] and followed it by one of his most revealing statements about the kind of poetry he intended to write: 'Je songe à une poésie qui serait de la psychologie dans une forme de rêve, avec des fleurs, du vent, des senteurs. D'inextricables symphonies avec une phrase (un sujet) mélodique, dont le dessin reparaît de temps en temps.'[8] The surprising thing is that, although this may not be how they are usually described, both 'L'Après-midi d'un faune' and a prose poem such as 'Frisson d'hiver' fit this description. And the same can be said in reverse, since some of Mallarmé's statements about poetry are surprisingly appropriate to Laforgue's late work.

The date of the four prose poems, 1864, was also the period during which Mallarmé was composing his major poem 'Hérodiade'. In order to describe the 'poétique très nouvelle' that it represented, he used a phrase with which Laforgue would have found himself in complete sympathy, since it equally well describes his own poetic aims and the achievement of the *Derniers Vers*: 'peindre non la chose, mais l'effet qu'elle produit'.[9] But this famous formulation is one that he could not have known, since it appeared in a private letter to Mallarmé's friend Cazalis. Thus, beneath dissimilarities of character, temperament, and

[4] *AH* 693. [5] *LA* 177. [6] Ibid. 178. [7] *AH* 757.
[8] Ibid. 757.
[9] Stéphane Mallarmé, *Œuvres complètes*, ed. G. Jean-Aubry (Paris, 1945), 1440.

execution, a similarity of poetic intention can be perceived. R. G. Cohn makes this interesting observation: 'it is possible and even likely that Laforgue, had he survived, would have developed the affinity [with Mallarmé] further: precisely because he was a very great artist, he could not have helped responding in some important ways to the challenge that Mallarmé represented to almost every leading man of letters of the twentieth century.'[10]

Although Laforgue had once met Mallarmé,[11] he must have seemed to him a somewhat distant deity. He sent him copies of both his published volumes (*Les Complaintes*, *L'Imitation de Notre-Dame la lune*) with some trepidation, and received from him sufficient praise to satisfy him ('le charmant billet de Mallarmé').[12] Yet it is clear from his fragmentary notes on him and from references in his correspondence that he grappled with central areas of Mallarméan poetics: the fusion of the intellectual and the plastic in his verse, for example ('ce n'est jamais une divagation d'images comme dans le rêve et l'extase inconsciente . . . mais de la divagation *raisonnée*. Toujours concret, jamais impalpable'[13]), and the poetic value of fractured sequentiality and of dislocated syntax ('quant à ses absences de syntaxe, ou plutôt de fil de phrase (principe Mallarmé). Je les adore parce qu'elles sont souvent adéquates à certains états et uniquement indispensables'[14]). And this even though he could not have known Mallarmé's correspondence, the early drafts of his poems, or his literary essays, most of which were in any case written later, and all of which seem today to shed invaluable light on Mallarmé's extraordinarily complex intellectual world.

Another example of the complications facing anyone attempting to assess Laforgue's knowledge of contemporary writings is that of Tristan Corbière. His *Amours jaunes* had been published as early as 1873, but Laforgue insisted that he did not know the work in 1885, when his *Complaintes* were published and were accused of resembling them;[15]

[10] R. G. Cohn, 'Laforgue and Mallarmé', in W. Ramsey (ed.), *Jules Laforgue: Essays on a Poet's Life and Work* (Carbondale and Edwardsville, Ill., 1969), 67.

[11] Henri de Régnier described the meeting at one of Mallarmé's 'mardis': 'il [Laforgue] répondait avec aisance et finesse aux propos de Mallarmé' (Régnier, *Faces et Profils: Souvenirs sur Villiers de l'Isle-Adam, Jules Laforgue, Mallarmé* (Paris, 1931), 34). After Laforgue's death, Mallarmé wrote: 'Quelqu'un était-il à l'enterrement du pauvre Laforgue; je n'ai été prévenu de rien, mais mène un deuil lointain et long.' (H. Mondor, *Vie de Mallarmé* (Paris, 1941), 516.)

[12] *LA* 145.

[13] *MP* 129.

[14] *LA* 68.

[15] 'Je n'ai lu du *Corbière* que dans les *poètes maudits*. Je n'ai jamais tenu les *Amours jaunes*. —Si, à ce propos, vous pouviez m'en avoir un exemplaire d'un prix pas trop fol, je vous serais bien obligé.' (J.-L. Debauve, *Laforgue en son temps* (Neuchâtel, 1972), 91.)

and, indeed, Verlaine wrote in 1883 that the volume was 'introuvable ou presque'.[16] On the other hand, Gustave Kahn reported that he had introduced Laforgue to it in 1880.[17] Laforgue never mentions Corbière in his correspondence between 1880 and 1885, while he refers to a number of other contemporary writers.[18] If Kahn did recommend the work to his friend, it had clearly made little impression at that point: 'On ne boit qu'aux sources qu'on connaît et qu'on désire', as Pierre Reboul says.[19] Laforgue was to drink at this spring later, as his provocative but perceptive notes on Corbière show,[20] emphasizing as they do, naturally enough in the circumstances, the ways in which the work of this least literary of poets differed from his own.

Some areas are less baffling. Heine was 'mon ami'.[21] Verlaine was an intimate kindred spirit: 'C'est bien celui dont je me rapproche le plus.'[22] Richepin interested him, but he soon outgrew him. He responded to Charles Cros's fantasy and wit; he had reviewed his *Coffret de Santal* already in 1879,[23] and continued to enjoy the work much later.[24] At the centre of the stage there was Bourget, to whose work he now seems to have been unjustifiably generous (*Les Aveux* is described as 'un livre de génie'[25]) partly perhaps in unconscious recognition of a personal debt to his mentor's friendship and kindness.[26] Lautréamont was completely unknown to him, as he was to all the young poets of his generation, even though the *Chants de Maldoror* had been composed as far back as 1868.[27]

16 Paul Verlaine, *Œuvres complètes*, iv (Paris, 1908), 7.

17 'Je lui révélais Corbière que je venais de découvrir . . . d'un de ses petits cousins qui . . . essayait de faire connaître *Les Amours jaunes* . . . Nous le trouvâmes admirable pour des raisons diverses.' (G. Kahn, *Les Origines du symbolisme* (Paris, 1936), 28.)

18 e.g. Gautier, Banville, Moréas, Marie Krysinska, Coppée, Villiers de l'Isle-Adam, Rollinat, Jules Lemaître.

19 P. Reboul, *Laforgue* (Paris, 1960), 37.

20 *MP* 119–28.

21 *AH* 687. This interesting area is fully explored by Ursula Franklin ('Reverberations of Heine in Laforgue's Lyrics', *Comparative Literature*, Spring 1987), who attributes Laforgue's obsessive image of the bleeding heart, his use of ballad and folk-song, his love of neologisms, and the development of his ironic tone to the influence of Heine, while she also points out some detailed instances of linguistic or thematic 'borrowings'. Convincing as these are, they do not, of course, rule out parallel French influences.

22 *AH* 886.

23 Jules Laforgue, *Les Pages de 'La Guêpe'*, ed. J.-L. Debauve (Paris, 1969), 164–6.

24 'Ce qui reste encore à côté de tout ça, c'est le *Coffret de Santal*.' (*LA* 92.)

25 *AH* 792.

26 'En laissant de côté l'inévitable subjectivité de tous mes jugements, il y a longtemps que je pense et dis à qui veut l'entendre que si quelqu'un a du génie parmi nos poètes, c'est Bourget.' (Ibid. 767.)

27 'Pendant toute la période symboliste, affirme Édouard Dujardin, je n'ai pas entendu prononcer une fois le nom de Lautréamont.' (I. D. Lautréamont, *Œuvres complètes*, (Paris, 1970), 24.)

Baudelaire was the poet from whom he probably learnt most, as did nearly all the other poets of the period, Verlaine and Mallarmé included. We shall consider his influence later. What is interesting at the moment is to observe how partial and subjective, although acute and arresting, Laforgue's vision of Baudelaire was even in 1885, when he wrote some preparatory notes for a study on him.[28] The best-known example of misapprehension is the unacceptable statement 'ni grand cœur, ni grand esprit',[29] which, however, is followed by one that cannot be faulted: 'mais quels nerfs plaintifs, quelles narines ouvertes à tout, quelle voix magique'.[30] One of the striking things about these nineteen pages of scribbled notes is the frequency of observations that apply just as much to himself as to Baudelaire. This is characteristic of the critical works of creative writers, and can of course be said of Baudelaire's own treatment of Hugo and Gautier. The following are merely some examples: 'il est un enfant en éveil, étonné, naïf';[31] Il a trouvé . . . le lyrisme plaintif—ses successeurs travaillent dans l'endolori':[32] 'Il n'a jamais cette vulgarité de la plupart des français jouer sur le pittoresque'.[33] According to Laforgue, it is Baudelaire who first describes 'la tristesse | misère du corps humain',[34] 'la femme sphinx malgré elle',[35] who writes 'des poésies . . . *sans sujet appréciable* . . . mais vagues et sans raison comme un battement d'éventail',[36] or who explores the following very Laforguian domains: 'l'artificiel, l'ironie, le paradoxe, l'excentrique, la volupté d'étonner, de déconcerter'.[37] Laforgue enjoys Baudelaire as a consumer. His criticism of him is occasionally uncomprehending, but more often brilliantly perceptive. He is receptive to the details of his use of language, to what separates Baudelaire from lesser poets, as he was immediately receptive to Rimbaud's stature as a poet (which Mallarmé never really seems to have been). Like the criticism of many writers, it reflects a vital relationship between the created object and the creative mind. Striving to be original, Laforgue admires the outstanding originality of this French 'esthète oriental'.[38] (In these few pages he applies the phrase 'le premier' to him no less than ten times.) It is nevertheless strange from a modern perspective to find that he does not quote at all

[28] J. A. Hiddleston's *Essai sur Laforgue et les* Derniers Vers *suivi de Laforgue et Baudelaire* (Lexington, 1980) contains the only complete transcription of these notes. I decipher 2 words on p. 103 differently. The differences are trifling, and I mention them only because they affect the sense: 'sa jupe *voir*' → 'sa jupe *noir[e]*'; '*on* appareille' → '*ou* appareille'. It is not unusual for Laforgue to omit agreements or misspell words in his rough notes. 'Sa jupe noir[e]' is clearly a reference to the widow described in 'A une passante'.

[29] Hiddleston, *Essai sur Laforgue*, 97.

[30] Ibid.

[31] Ibid. 107.

[32] Ibid. 99.

[33] Ibid. 107.

[34] Ibid. 95.

[35] Ibid. 94.

[36] Ibid. 104.

[37] Ibid. 96.

[38] Ibid.

from what now seems one of Baudelaire's major innovations, the prose poetry of *Le Spleen de Paris*. A simple explanation may well be that he did not possess a copy of these poems, but their absence must also reflect the literary moment. Baudelaire's achievement in the prose poems was not yet fully recognized, and his reputation, even for young poets bent on innovation, rested less on the 'miracle'[39] of his fusion of poetry and prose than on his 'voix magique' in *Les Fleurs du mal*.

If Baudelaire was 'un vrai Dieu'[40] for others apart from Rimbaud and Laforgue, the poetic scene in Paris was dominated by a group who called themselves 'Hydropathes', most of whom are almost completely forgotten today. Their leader was a bohemian figure, Émile Goudeau, and his followers represented, in his words, 'un inextricable fouillis de tendances diverses, contraires'.[41] Their numbers grew swiftly from thirty to some three hundred. They met twice a week in a series of halls in Paris, and recited poetry aloud. For these young 'décadents', as they also termed themselves, Naturalism and aestheticism rubbed shoulders. They admired equally Huysmans's Naturalist novel *En ménage* and its decadent successor *A rebours*. They were devotees of both Baudelaire and Taine. Placed between the formal rigour of the Parnassians and the anarchy of early Symbolists—later styled by Verlaine 'cymbalistes'—the Hydropathes combined sincerity with humour, and aimed to 'faire retourner l'inspiration poétique au sens de la mesure et de la vérité'.[42] This is probably how one should picture them—and the young Laforgue: 'On imagine les voix vibrantes jetant les vers en d'éclatantes sonorités devant un auditoire en grande partie universitaire, c'est-à-dire instruit et gai, ayant fui après la journée d'étude, la chambre maussade de l'hôtel . . . pour accourir devant le Tréteau des Muses.'[43] Much later, Bourget, one of their original number, described them as 'des observateurs, des artistes soucieux des réalités vivantes et rebelles au pessimisme de la philosophie étrangère qui tendait à imprégner toute la littérature'.[44] Laforgue was, as we shall see, highly receptive to that 'foreign' philosophy, but he allied his interest in it with a humour that owed something to the 'groupe d'hommes de talent, plaçant leur inspira-

[39] 'Quel est celui de nous qui n'a pas dans ses jours d'ambition, rêvé le miracle d'une prose poétique . . .?': from Baudelaire's *dédicace* to Arsène Houssaye in *Le Spleen de Paris*.

[40] A. Rimbaud, 'Lettre du voyant' to Paul Demeny, in *Œuvres complètes*, ed. R. de Renéville and J. Mouquet (Paris, 1951), 257.

[41] R. de Casteras, *Avant le chat-noir les Hydropathes* (Paris, 1945), 53.

[42] Ibid. 238.

[43] Ibid. 71–2.

[44] Ibid. 239.

tion sous le signe de la fantaisie, du persiflage, de l'ironie suraiguë'.[45] To realize how far removed they were from Symbolism as it developed, one has only to consider Mallarmé's reaction in 1886 to the literary journal that promoted them. While permitting two of his 'petits poèmes' to be printed in it, he wrote: 'mais quel titre abominable que *La Décadence* et comme il serait temps de renoncer à tout ce qui y ressemble!'[46] Kahn, in *Les Origines du symbolisme*, gives a clearer idea of their importance, describing them as 'un énorme groupe . . . qui englobait . . . tous les méconnus, tous les passionnés d'écriture artiste, ou plutôt d'écriture expressive et de forme nouvelle.'[47]

This, then, was the scene that confronted the 18-year-old Laforgue when he appeared in Paris in 1879. This study aims to trace his growth as a poet from this point. We shall be concerned both with the writers who influenced him and with the highly individual path that he took himself. We shall consider his ancestry among philosophers and cosmic poets in his first and most derivative volume (the unpublished *Le Sanglot de la terre*), the ironic exploitation of the folk-song in the second (*Les Complaintes*), the use he made of the Pierrot 'mask' in the third (*L'Imitation de Notre-Dame la lune*), the very profound reorientation that the next unpublished volume represented (*Des Fleurs de bonne volonté*), and, finally, his move towards his major achievement, the posthumously named *Derniers Vers*, complex monologues in free verse (of which he has the strongest claim to be the inventor), separate poems in an incomplete volume, cut short by his death.

Influences include those of other art-forms apart from poetry, in particular that of Impressionist painting and of the fashionable 'Wagnerism'. They include also the effect on his poetic style of translating Walt Whitman's free verse into French, and of writing in prose the short stories that make up the *Moralités légendaires*. We shall observe him closely at work (particularly in the variants of the *Fleurs*), and discover how deceptive was his appearance of negligent ease and also how flexible was his method of composition, relying on the constant interaction of intellect and intuition, analysis and fantasy. We shall compare him with contemporaries such as Gustave Kahn and with a later poet, the young T. S. Eliot, who was, as is well known, strongly influenced by his work, and see what these comparisons tell us about Laforgue himself. We shall attempt to assess the interest of his work at each of the varied

45 Ibid. 29.

46 Stéphane Mallarmé, *Correspondance*, iii, ed. H. Mondor (Paris, 1969), 62.

47 G. Kahn, *Symbolistes et décadents* (Paris, 1902), 57–8.

stages of his brief career. Richard Ellmann said of James Joyce that we are still learning to be his contemporaries. Laforgue was both of his time and strikingly ahead of it. 'Il faut être de son âge et même en avant de son âge', he wrote, 'sous peine de passer pour un Monsieur antédiluvien et d'être recommandé aux paléontologues de la critique littéraire.'[48]

In considering Laforgue's origins, one must remember, not the history of literature as we have reconstructed it since his death, but the writing that had actually received recognition and that had been published when he started to write himself. However, whether we consider him in his own terms or in ours, we will be struck by his restless originality, rising from and beyond the situation of 1880. '. . . la destinée d'un artiste', he wrote, 'est de s'enthousiasmer et se dégoûter d'idéaux successifs.'[49] It is the resulting literary innovation that I wish to explore. It involved necessary and valuable disruption. 'If we think,' wrote Edgar Wind, 'for example, of Manet, Mallarmé, Joyce or Stravinsky, it would seem that all the artistic triumphs of the last hundred years were in the first instance triumphs of disruption: the greatness of an artist became manifest in his power to break up our perceptual habits and disclose new ranges of sensibility.'[50] This was precisely what Laforgue may be said to have achieved, and it was the result, at one level, of a desire to see things without cant or prejudice. 'Quand est-ce que nous nous montrerons adéquats à la *valeur des phénomènes*,' he asked, 'et vivrons-nous *justes de ton*?' (Laforgue's emphases).[51] The question he consistently asked of literature was a parallel one: 'Quand écrirons-nous justes de ton?'

[48] *AH* 751.
[49] Ibid. 822.
[50] E. Wind, *Art and Anarchy* (London, 1985), 18.
[51] *MP* 15.

The Life

LAFORGUE died of tuberculosis on 20 August 1887, when he had just reached the age of 27. By this time he had written a great deal. Unpublished juvenilia include a verse comedy (*Tessa*, 1877), a short novel (*Stéphane Vassiliew*, 1881), and parts of another, of which we have only fragments (*Un raté*). There was highly original art criticism, particularly that dealing with Impressionism, and remarkable notes for essays on Baudelaire and Corbière, and there was a detailed book describing the court and the town of Berlin (*Berlin: La cour et la ville*). His most successful work apart from his poetry was a volume of parodic short stories, the *Moralités légendaires*, which were not published until late 1887.

His poetry consisted of five quite separate volumes of verse, only two of which were published during his lifetime: *Les Complaintes* (1885), and *L'Imitation de Notre-Dame la lune* (1886). Eleven of the twelve substantial poems that make up the posthumous *Derniers Vers* were, however, published during his lifetime, all except one in pairs in the journal *La Vogue*, as was a group of poems from the projected volume *Des Fleurs de bonne volonté* under the title *Le Concile féerique*. His starting-point was the set of poems for the volume *Le Sanglot de la terre*, which he soon rejected. Before we consider the aesthetic journey that these productions represent, a brief outline of his life may be helpful.

Jules Laforgue was born on 16 August 1860 in Montevideo, the second son of Charles Benoît Laforgue, a schoolteacher who came from Tarbes, and of Pauline Lacolley, who was also French in origin. The aim of the Laforgues was to return to France when their financial situation allowed it, and in 1866 they thought this time had come. They set sail—mother, five children, and two maternal grandparents—for Bordeaux, a journey that took seventy-five days and one that Jules was never to forget. They went to Tarbes, where they had cousins and where Charles Laforgue joined them as soon as he had settled their financial affairs in Uruguay. By 1869 he found himself unable to support his ever-growing family in France, and he returned with them to Montevideo, leaving his two eldest sons, Émile and Jules, as boarders at the *lycée* in Tarbes, where Jules remained until 1876. The holidays were spent with their father's cousins, the Darrés, whom the Laforgue children found strict and unfriendly. Jules, a gentle and dreamy child, appears to have had little academic success at school. He shone only in religious education and drawing, but he edited a school magazine with Émile, and, when he left school, contributed articles to the Toulousian literary review *La Guêpe*, which was produced by former school-friends.

In 1875 the Laforgue family returned to Tarbes, where, in November, an eleventh child was born. The following year they moved to the outskirts of Paris, 66 rue des Moines, in the Batignolles area, apparently for the sake of Émile's and Jules's education, for their father had intellectual ambitions for his sons. Jules went to the Lycée Fontanès (later Condorcet), where he prepared for the *baccalauréat*, but, although he once did well in the initial written examination, he floundered in the oral and in fact never obtained the qualification. On 6 April 1877 Pauline Laforgue died, having shortly before given birth to a premature child who did not survive. She was 39. Jules's sense of grief and loss is seen in his earliest poems. Later he was to give the, no doubt, equally accurate reverse side of the picture: 'J'avais presque pas connu ma mère.'[1]

The remaining family moved in 1879 to a more central Paris address, 5 rue Berthollet, but soon after returned to Tarbes because of the illness of Charles Laforgue. Jules and Marie, the eldest daughter, to whom Laforgue was deeply attached, stayed on alone in Paris. (Émile was doing his military service.) Since he had not passed his *baccalauréat* and so could not become a university student, Jules joined the studio of Henri Lehmann at the École des beaux-arts (where Émile was inscribed), and attended the lecture courses of Hippolyte Taine. He frequented libraries and museums and read widely. He formed friendships that were to last the rest of his life, with the poet Gustave Kahn and with Charles Henry, whose wide-ranging interests included mathematics, neurology, aesthetics, and philosophy. He introduced himself to the poet and novelist Paul Bourget, who became a loyal friend and mentor. Ironically, he told Laforgue he must think for himself, but quite failed to understand him when he did.

In 1881 Bourget found him a post as secretary to the art historian and collector Charles Ephrussi, whose volume *Albert Dürer et ses dessins* Laforgue helped to prepare. He owed his early initiation into the world of art criticism to this equally loyal friend and supporter. In September 1881 Marie returned to Tarbes to look after their father, who was dying of tuberculosis, and Jules moved to an attic room in 43 rue Monsieur-le-Prince. Bourget and Ephrussi came to his rescue, finding him a post as

[1] 'Avertissement', *Des Fleurs de bonne volonté.*

French reader to the Empress Augusta in Berlin, a remarkable elevation for the impecunious young poet. He was on the point of leaving when his father died on 18 November 1881. His letters to Marie reveal, alongside the expected grief, extreme concern that he had not been able to prove himself to his father, and a later poem suggests a somewhat distant relationship: 'Mon père, un dur par timidité'.[2]

His duties at the court of the Empress were to read to her, once or twice daily, excerpts from the French press and selections from novels, biographies, and poems of his choosing, all of which he did to her evident satisfaction. He earned a good salary (9,000 francs a year), lived in a luxurious appartment in an atmosphere of elegance, had leisure to read and write—and clearly did read widely, while he wrote intermittently with a frenzied enthusiasm, between spells of ennui and loneliness.

He followed the court on its routine journeys to a series of watering-places: Baden-Baden, Koblenz, and Potsdam, he was a frequent visitor to the opera, and he returned to France for vacations when he could afford to travel there. He kept in touch with his Parisian friends by frequent letters. He made new and close friendships, with the journalist Théodore Lindenlaub, and particularly with the younger of the two Ysaÿe brothers, Théophile. (The elder, Eugène, was to become the famous violinist.) He met figures of the French musical and literary worlds, such as Édouard Dujardin and Téodor de Wyzewa, when they visited Berlin, and clearly impressed them. The poor young poet who, in Marie's words, 'avait souvent les habits râpés',[3] had become an elegant and sophisticated aesthete.

His diary of 1883 reports in brief and enigmatic phrases a relationship with a woman who is always referred to as 'R'. The most likely attribution, since she followed exactly the movements of the court, is a lady of the Bedchamber, Garderobenfrau Rosalie Marie Hélène Bachem, second daughter of Alexander Bachem, the Mayor first of Koblenz and then of Cologne. The surprising fact—if, as seems likely, she was the mysterious 'R'—is that she was twenty years Laforgue's senior. The affair was tempestuous. At one point, which fits our supposition, Laforgue noted: 'Elle était née pour être mère'[4], but in fact she never even married.

[2] Ibid.

[3] A private letter to Georges Jean-Aubry, dated 22 Mar. 1922, now in the Bibliothèque Jacques Doucet.

[4] *AH* 882.

She could presumably scarcely have been more different from Leah Lee, the young English girl whom Laforgue married. Like him, Leah was a foreigner, living partly by giving English lessons in Berlin, partly on an allowance from her father. She taught Laforgue English, and they read *Hamlet* together. Like him, she had no family to return to. On her mother's death her father had married his children's governess, and the older children had all left home. Laforgue proposed to Leah on 6 September 1886, after a period of doubt and anxiety that was more than justified. He could not remain in the service of the Empress if he was married, and he knew this. But he suffered from his literary isolation in Germany, and wished to return to Paris. Sadly, both the dream of living off his pen and of being able to support a wife were hopelessly unrealistic. The couple were married in London, in St Barnabas Church, Addison Road, on 31 December 1886, the same church in which T. S. Eliot married his second wife. They moved to Paris to live in a fifth-floor flat at 8 rue de Commaille.

From the start the young couple's lives were dogged by ill health and poverty. Few of his friends suspected, and even Laforgue did not fully realize, the seriousness of his condition. Nine months after his death in 1887, Leah died, also of tuberculosis, on 6 June 1888. Although Laforgue died first, it may be that he had contracted the disease from her. Just before his death, Laforgue set the literary record straight in an untypical statement to Marie:

> il y a longtemps que tu ne sais plus rien de mes affaires littéraires. Ce serait trop long à détailler, mais sache d'un mot que j'ai le droit d'être fier; il n'y a pas un littérateur de ma génération à qui on promette un pareil avenir. Tu dois penser qu'il n'y a pas beaucoup de littérateurs qui s'entendent dire: 'Vous avez du génie.' Hélas! qu'il me tarde d'être guéri . . . [5]

[5] *OC*, v. 196. David Arkell's valuable biography, *Looking for Laforgue: An Informal Biography* (Manchester, 1979), contains much original research. The riddle of Leah Lee's identity, caused by the inaccurate record of her age on her marriage certificate, was, however, solved many years before Arkell undertook the same journey, by R. R. Bolgar (see his letter to the *Times Literary Supplement*, 5 Feb. 1954, and A. Holmes, 'The Poetic Development of Jules Laforgue', Ph.D. thesis (Cambridge, 1956)).

1

Le Sanglot de la terre: 'La note aiguë de ma froide, froide destinée'

DERIVATIONS: COSMIC POETRY AND HARTMANN

'SI j'avais des antécédents à un point quelconque de l'histoire de France! | Mais non, rien',[1] was Rimbaud's lament. Laforgue, equally an innovator, could not have made a statement of this kind. He admired Rimbaud's poetry for being 'sans attaches',[2] perhaps all the more because his own was unmistakably intertextual. In early work such as the poetry he wrote for *Le Sanglot de la terre* one is reminded of the writings of others by a stream of virtually unconscious echoes, as is common in young poets. In later works he pointed out predecessors and contemporaries himself, using them as vehicles for irony. Self-conscious intertextuality was not limited to references to other poets: he parodied the language of his preferred philosophers, Schopenhauer and Hartmann; he adapted well-known folk-songs in the *Complaintes*; he rewrote legends in the *Moralités légendaires*. Down to his reuse of his own earlier verse, his is 'quotational' poetry, a fact that he drew attention to himself: 'Mon cœur est un lexique où cent littératures | Se lardent sans répit de divines ratures.'[3]

Despite this fact, no poet plays for him the role that he played for the young Eliot—that is, reveals to the later poet the direction that will be particularly suited to him. Predecessors provide both his early themes and vocabulary, but, initially, they seem if anything to lead him astray. That at least is one's impression from *Le Sanglot de la terre*, the plan of which consists of five sections in the manner of forerunners such as Baudelaire: 'Lamasabacktani'; 'Angoisses'; 'Poèmes de la mort'; 'Résignations infinies'; 'Spleen'. The volume was to be 'l'histoire, le journal d'un parisien de 1880, qui souffre, doute et arrive au néant'.[4] The personal solitude and loss of religious faith that pierce through the rhetoric of the poems are based on experience, and Laforgue convinces us of the reality of both, without succeeding in making them artistically interesting. This is partly because he uses the methods of his elders

[1] Arthur Rimbaud, 'Mauvais Sang', *Une Saison en enfer*.
[2] *LA* 187.
[3] 'Complainte-litanies de mon sacré-cœur', *Les Complaintes*.
[4] *MP* 8.

even when his conscious intention is to differ from them. It is also because of a blurring of the real and experienced with the borrowed and, by now, trite. What are we to make of a 20-year-old poet who refers to himself as if he were the mature Hugo, or Vigny, or the disillusioned Leconte de Lisle,

> Lâche j'ai vu partir l'Art ma dernière idole,
> Le Beau ne m'étreint plus d'un immortel transport,
> Je sens que j'ai perdu, car avec l'Art s'envole
> Cette extase où parfois le vieux désir s'endort.
> ('Citerne tarie')

or a world-weary Baudelaire?

> Je ne vous aime pas, non, je n'aime personne,
> L'Art, le Spleen, la Douleur sont mes seules amours;
> Puis, mon cœur est trop vieux pour fleurir comme aux jours
> Où vous eussiez été mon unique madone.
> ('Excuse mélancolique')

Laforgue adopts the language of age, which he uses indiscriminately to refer to himself and to an overripe society: 'Trente siècles d'ennui pèsent sur mon épaule.'[5] The 'immenses lassitudes'[6] that he suffers from are less convincing as a psychological condition than as a reflection of general historical circumstances, the French political and intellectual situation in 1880, ten years after the débâcle of 1870 and the passing of an age of obvious poetic achievement.[7] They highlight the nature of Laforgue's literary problem: how to follow writers who gave lyrical poetry a metaphysical direction—Hugo, Vigny, Nerval, Baudelaire—when their lesser successors—Leconte de Lisle, Sully Prudhomme, Louise Ackermann, Cazalis (to name only the few that Laforgue mentions in connection with this volume)—had seemed to exhaust the range of possible responses to the threat of the godless universe.

Laforgue compared himself with the latter group, and found each poet wanting: 'Leconte de Lisle pas assez humain . . . Cazalis trop dilettante, Mme Ackermann pas assez artiste, pas assez fouillée, Sully Prudhomme trop froid, trop technique.'[8] The dilettantism that Laforgue

[5] 'Citerne tarie', *AH* 269.
[6] 'Éponge pourrie', *AH* 271.
[7] Reboul describes the situation succinctly: 'Laforgue fut le témoin dolent et magnifique d'une brève époque où, entre l'Empire effondré, la Commune explosée en une horrible *Götterdämmerung* et la République encore instable, entre une bourgeoisie "parvenue", et les alibis d'un symbolisme incertain, le désespoir des jeunes gens prenait la forme vaine et digne d'une rébellion artistique, appuyée sur un scientisme qu'elle dépassait.' (P. Reboul, *Laforgue* (Paris, 1960), 200.)
[8] *MP* 8.

was later to adopt seemed at this stage unsuited to the seriousness of his subject-matter. On the other hand, a 'langue d'artiste'[9] was already essential, and, though he wrote this volume, as he was later to say, without possessing 'un brin de mon métier',[10] he was quick to spot its absence in others. Leconte de Lisle and Prudhomme lacked feeling, and this is essentially what he intended to supply. His poetry was to be intense: 'Je n'écris pas une phrase, un vers sans vouloir du suraigu.'[11] Rather charmingly, he registered amazement that philosophers, who daily demolished false ideals of justice, morality, or religion, were 'si peu émus'.[12]

But the pursuit of intensity led to a type of eloquence that he had rejected by the spring of 1882: 'à cette époque je voulais être éloquent, et cela me donne aujourd'hui sur les nerfs.—Faire de l'éloquence me semble si mauvais goût, si jobard!'[13] He now found the poems 'mauvais manifestement',[14] but did not consider either his subject-matter or this stage in his development insignificant. It was his literary method that was at fault. After he had moved to a totally different aesthetic for his *Complaintes*, he insisted, against the advice of his friends, on opening the volume with a long introductory poem, the 'Préludes autobiographiques', because 'ce fut une étape et je tiens à dire (aux quelques à qui j'enverrai le volume) qu'avant d'être dilettante et pierrot j'ai séjourné dans le Cosmique.'[15] He mocked it, of course, in the poem's opening lines:

> En voulant mettre un peu ordre dans ce tiroir,
> Je me suis perdu par mes grands vingt ans, ce soir
> De Noël gras.
> Ah! dérisoire créature!

At the time, however, the 'note aiguë de ma froide, froide destinée'[16] was real enough. Laforgue had reached a state of unbelief that went far beyond the religious doubts of Hugo, Nerval, or Vigny. The situation that the volume describes is blurred more because of the mixture of the sincere and the borrowed than because of any real uncertainty of philosophical stance. Laforgue's questioning of the universe, when it occurs, lacks the intensity of Hugo's attempt to pierce the 'walls' of infinity, and consequently invents none of the extraordinary metaphors fusing concrete and abstract that abound in Hugo's apocalyptic poetry.[17]

[9] Ibid. [10] *AH* 763. [11] Ibid. 741. [12] *MP* 7.
[13] *AH* 782. [14] *LA* 79 [15] Ibid. [16] *AH* 763.
[17] 'Le Hibou' provides some splendid examples.

Laforgue's cosmos remains conventional when it, rather than the poet, is the focus of the poem. And yet its importance to him is undeniable. When, in his later work, it becomes an objective correlative[18] for the poet, rather than being the directly addressed or described subject; when the stars, sun, moon, wind, and rain echo and modulate human emotions, the cosmos has taken its proper place as an element in the shifting human consciousness. Man's isolation in an unknowable universe is never again treated by Laforgue as a main philosophical theme; instead it becomes an integral element of the narrator's psychology.

Already in this volume the direction of cosmic poetry has, to some extent, been reversed. This is because, in Christian terms, Laforgue had quite simply given up the quest, and, despite verbal echoes, his intellectual climate is equally removed from Vigny's idealization of the intellect and of a poetry of ideas illuminating man's future upward path. Laforgue's celebration of the unconscious force driving the universe, against which the human mind is powerless, inevitably leads away from the rational principle and back to the deeper, unknowable regions of the psyche. It is logical, then, even if it is ironical, that the emotions, not the rational faculties, should dominate in this supposedly philosophical volume; equally, that this 'cosmic' poetry should interest us less for its depiction of the universe than for its most personal metaphor, that of the bleeding human heart.

Laforgue's private jottings give the most useful insight into the view of life that he held and wished to preach—because he saw his role as prophetic. While he admitted that there was 'pas de remède absolu, universel, qui supprime le mal universel', he proposed two solutions: 'Bouddha, L'Inde vénérable—Shopenhaüer [*sic*], Hartmann'.[19] The aim of both was 'le repos', brought about by distancing oneself from the evils and sufferings of the universe. An earlier stage of immersion in these sufferings was necessary:

> Avant d'arriver au renoncement, il faut souffrir au moins deux ans: jeûner, souffrir de la continence, saigner de pitié et d'amour universel, visiter les hôpitaux, toutes les maladies hideuses ou tristes, toutes les saletés, se pénétrer de l'histoire générale et minutieuse, en se disant que cela est réel . . . la lire avec *sympathie* (le premier don du sage) comme Carlyle ou Michelet.[20]

One is reminded of Flaubert's statement concerning the artist's distance from the vulgar emotions, which also elevates the role of sympathy: 'Je ne veux avoir ni amour, ni haine. Quant à la sympathie . . . jamais on

[18] See Ch. 8. [19] *MP* 10. [20] Ibid. 11–12.

n'en a assez.'[21] By renunciation, a state beyond man's personal and ephemeral misery is reached. 'On est affranchi pour un instant du temps', Laforgue says in a Proustian phrase, 'on meurt à la conscience de son individualité.'[22] Individual will, which is inseparable from suffering, is destroyed, and a state of contemplation is attained that may be aesthetic, scientific, or philosophical. Illustrating his keen interest in both these areas at the time, he adds: 'ces deux dernières sont les plus sûres de quiétude'.[23] He sets religions that offer the vision of an after-world as an opiate to human fears firmly in the past: 'Le grand bienfait des religions est d'avoir bercé la douleur et les effrois de l'homme d'un au delà d'éternelle douceur. Aujourd'hui, qu'on y fasse attention, cet espoir s'en va, on ne croit plus.'[24] All these ideas, heavily dependent on Schopenhauer and Hartmann, as well as on his adolescent attachment to Buddhism, were to be set out in the 'Bible nouvelle', presumably a prose work separate from the *Sanglot*. In his notes he ended his outline of the *Sanglot* with the section 'Résignations', appropriate for the final point of the metaphysical journey. In his detailed plan for the volume, however, he reversed the order of the last two sections, and gave the final emphasis to the aesthetically more successful 'Spleen' poems, which represent the metaphysical failure to reach, beyond individual suffering, to a contemplative state, and stress the personal rather than the metaphysical.

At 20 years of age, Laforgue was in the characteristically modern position of a would-be writer for whom religion had no attractions, who scorned reason, who had no clear literary guide, who trusted only the transient emotions, but who was at the same time considerably attracted by a few grand philosophical panoramas that had an influence on the development of his ideas. His earliest writings refer to Schopenhauer—as to Hegel—with familiarity.[25] Eduard von Hartmann, who is first mentioned in a letter of December 1880 to Gustave Kahn,[26] came later, and was for Laforgue both a comforting and a liberating presence: comforting, because he provided a world-view that, pessimistic though it was, amounted to a system (his *Philosophie des Unbewussten* had been translated into French by 1877, and became Laforgue's *livre de chevet* and his new Bible): liberating, because Hartmann set art firmly in the realm of the unconscious, beyond the power of human reason, and, for a poet who needed to break new aesthetic ground, this was a kind of

21 Gustave Flaubert, *Correspondance*, iii. ed. J. Bruneau (Paris, 1991), 786.
22 *MP* 11. 23 Ibid. 24 Ibid. 10. 25 *AH* 236.
26 Ibid. 687.

salvation. Already in the *Sanglot* the unconscious licensed a number of linguistic freedoms. The fact that the unconscious was part of the psychological rather than the strictly philosophical realm was also vitally important. Laforgue had little that was original to offer in philosophical terms, although he became a sophisticated commentator on other philosophers, as can be seen from his notes for an essay on Schelling:[27] his artistic gifts were clearly those dependent on an acute sensibility.

Hartmann's outline of the three stages of human illusion (the first that happiness can be attained by the individual in this life, the second that it can be attained in an afterlife, and the third that it will be attained at some future state of the world's evolution) are referred to when Laforgue describes the plan of his novel *Un raté*,[28] and would have been perfectly acceptable to him in 1880. But one of the reasons that initially attracted him to Hartmann was no doubt that he solved for him the problem of the gulf between idealist and determinist aesthetic positions by putting forward a view that combined them.[29] Laforgue was in sympathy with the idealism of a philosopher such as Hegel, and was ready to believe that the universe was governed by an idea, a Schopenhauerian will or an unconscious force. But he also believed in evolving ideals of beauty, since for him the essential characteristic of art was its reflection of the ephemeral and the contingent. Hartmann discovered a path between the entrenched opposing positions by means of his proposition that a constantly evolving unconscious force was the ultimate metaphysical reality. This force was supported, but only in auxiliary fashion, by a rational principle. Hartmann's separation of talent from genius would have found a sympathetic ear in Laforgue, and Laforgue's frequent use of the word 'génie' must be in part an echo of Hartmann. Talent was linked with the conscious intelligence; genius—and inspiration, which was a necessary adjunct of genius—with the irrational,

[27] Unpublished. Formerly in the Jean-Aubry collection.

[28] *AH* 687.

[29] Eduard von Hartmann, *Philosophie de l'inconscient* (Paris, 1877), 298 (according to R. MacKenzie, this is the first recorded use of 'inconscient' as a noun: see *Modernism and the European Unconscious*, ed. P. Collier and J. Davies (Oxford 1990), 151): 'Les empiriques ont raison d'affirmer que tout jugement esthétique doit dépendre de conditions psychologiques et physiologiques . . . Les idéalistes ont raison de soutenir contre leurs adversaires que le jugement esthétique se fait en dehors de la conscience. . . . Mais ils ont tort, à leur tour, de nier l'activité propre de la faculté esthétique a priori, en affirmant que l'idéal est fixé une fois pour toutes . . . Le lien objectif de la faculté esthétique avec les autres facultés de l'esprit demeure ainsi éternellement incompréhensible; et le caractère immuable de l'idéal ne peut se soutenir en regard de la diversité infinie des cas particuliers qu'il doit expliquer.'

unknowable because 'unconscious' regions of the human mind. The progeny, with deviations, of course, of this line of thought include Proust and even Valéry. Thomas Hardy was a devotee of Hartmann, and Freud an interested reader. At every stage in artistic composition—even when dealing with detail rather than with initial conception—the artist must follow his instinct. Hartmann quotes a letter of Mozart's in which he explains reluctantly, because he says the process is mysterious and defies explanation, how he arrived at his musical compositions. The letter is as extreme an illustration as one could hope to find of unconscious creativity, and perhaps not many poets could claim to hold the entire work of art in their minds as Mozart does his composition:

les pensées me viennent en foule et le plus aisément du monde. D'où et comment m'arrivent-elles? Je n'en sais rien, je n'y suis pour rien . . . Mon âme s'enflamme . . . L'œuvre grandit, je l'étends toujours et la rends de plus en plus distincte; et la composition finit par être tout entière achevée dans ma tête, bien qu'elle soit longue. Je l'embrasse ensuite d'un seul coup d'œil, comme un beau tableau ou un joli garçon. . . . Tout cela, l'invention et l'exécution, se produit en moi comme dans un beau songe très distinct.[30]

Even if the process is described here in its ideal form, it is easy to believe that something analogous to this must be at the root of every work of true originality. For Laforgue, the relegation of the rational principle to second place, and the insistence on listening to the inner unknowable voice that dictated even the 'details' (*l'exécution*[31]) of composition were vitally important. Later, but not yet here, the variants of his poems reveal how ready he was to abandon logic and to allow instinct to lead him. In 1885 he wrote some notes that describe the unconscious in a revealing manner. (They mingle first- and third-person narration, and were probably rough drafts for a novel figuring a Laforguian hero.) We see the unconscious, as we would expect, both as an external force to which man must submit and as a personal 'unknown', a source of infinite richness:

Je me sens si pauvre si connu tel que je me connais moi, Laforgue, en relation avec le monde extérieur—Et j'ai des mines riches, des gisements, des mondes sous-marins qui fermentent inconnus—Ah! c'est là que je voudrais vivre, c'est là

30 This is the French translation that Laforgue would have read: Hartmann, *Philosophie*, 308-9.

31 Some critical confusion has resulted from neglect of Hartmann's insistence that 'exécution' as well as 'invention' stems from the unconscious.

que je voudrais mourir . . . Aujourd'hui tout préconise et tout se précipite à la culture exclusive de la Raison, de la logique, de la conscience—

La culture bénie de l'avenir est la déculture, la mise en jachère.

Nous allons à la dessication: squelettes de cuir, à lunettes, rationalistes, anatomiques.

Retournons mes frères vers les gdes eaux de l'Inconscient.[32]

Laforgue was to echo Mozart's description much earlier, in 1882, when he described his poem 'Sieste éternelle' as 'un autre sonnet, sorti je ne sais d'où de mon être et je ne sais comme'.[33]

THE BEGINNINGS OF A PERSONAL STYLE

Le Sanglot de la terre (1880–2), Laforgue's first volume of poetry, reveals, as one would expect, the conscious and unconscious debts of a young poet to his immediate predecessors. Laforgue's important poetry is to come later, but it is interesting to observe how, at this stage, he was beginning both to use and to reject his models, moving towards the personal and original style that he was eventually to create.

Pierre Reboul's journey through Laforgue's borrowings from contemporaries[34] (Hugo, Vigny, Gautier, Baudelaire, Cros, Bourget, Banville, early Verlaine, Heine in the fashionable translation by Mérat and Valade) can leave no doubt of the large number of, presumably, chiefly unconscious debts in this volume. Sensitive to, and capable of responding to, the climate of the time—to its pessimism, its Naturalism, and its decadence[35]—Laforgue clearly wrote without realizing how much his style was formed by others. Some of the most carefully worked poems in the volume were taken to his mentor Paul Bourget,[36] and he is one of the writers whose influence is most visible. There can be no question of parody here. Nor can Bourget have seen imitation in the poems to quite the extent that we do. Pastiche becomes more obvious with the passage of time. Mallarmé's and Verlaine's early

[32] *FV* 104–5. [33] *AH* 788. [34] Reboul, *Laforgue*, 8–57.

[35] Reboul gives an excellent account of the intellectual climate of 1880 (ibid. 10–29).

[36] Bourget clearly recognized that Laforgue had still to learn to think for himself: 'Je me souviens du temps où je portais à Bourget des pièces de théâtre, des chapitres de roman, et des masses de vers, en songeant: de ce coup-ci il va être épaté! Et il me répondait le dimanche suivant: "Vous ne savez pas encore le français, ni le métier du vers, et vous n'en êtes pas encore à penser par vous-même"'. (*OC*, v. 146.) At this stage of his development there can be little doubt that Laforgue learnt a great deal from Bourget, most valuably perhaps 'le métier du vers'.

Baudelairean sonnets are an example. It is partly that increased fame leads to easy recognition of the briefest phrases. But it is also that imitated elements stand out in an unnatural isolation once the linguistic manner of the day, of which they formed a part, has become archaic. Laforgue has only to write in his sonnet 'La Première Nuit', the first of the 'Poèmes de la mort': 'Voici venir le soir doux au vieillard lubrique', and 'C'est l'heure où l'enfant prie', for a modern reader to link the poem with Baudelaire's 'Crépuscule du soir', even though only three words are repeated in each case. (Baudelaire wrote: 'Voici le soir charmant, ami du criminel', and 'C'est l'heure où les douleurs des malades s'aigrissent'.) The manuscript of 'Recueillement du soir', another poem where Laforgue uses these phrases, has the abbreviation 'Baud' scribbled by the title, showing his awareness that the title, and perhaps the poem, echoed Baudelaire.[37]

'Recueillement du soir', a 'grosse pièce' (Laforgue divided his volume into sonnets and 'grosses pièces'), is more expansive than 'La Première Nuit'. Baudelairean types abound—beggars, blind men, prostitutes, criminals, workers, the dying, the mad, the monk—amid a range of milieux from cafés to hospitals. At the very least the poem reveals Laforgue's admiration for the poet who had first opened up the modern city to poetic treatment. In his notes on Baudelaire Laforgue praised the accuracy of descriptions of 'les becs de gaz que tourmente le vent, la Prostitution qui s'allume dans les rues, les restaurants et leurs soupiraux, les hôpitaux, le jeu, le bois qu'on scie en bûches qui retentissent sur le pavé des cours, et le coin du feu, et les chats'.[38] Laforgue's subject was also the milieu in which the modern Parisian suffered with its 'couchants, la Seine, les averses, les pavés gras, les Jablochkoff'.[39] His poetry, just as much as Baudelaire's, was dependent on capturing the moment. For Baudelaire, the ephemeral was the necessary path to the absolute: for Laforgue, it was the substance of art itself.

He may well have seen his Parisian poems as variations on Baudelaire's.[40] The section 'Poèmes de la mort' had an alternative title of

[37] 'Recueillement', 'Harmonie du soir'. [38] *MP* 111. [39] Ibid. 8.

[40] In *Semiotics of Poetry* (London, 1980) M. Riffaterre analyses 'Sieste éternelle', a sonnet that is clearly influenced by Baudelaire, from the angle of intertextual semiotics. The poem, which Laforgue did not attempt to publish, was sent in July 1882 to the poetess Mme Mültzer, whose salon Laforgue had attended in Paris and with whom he kept up a flirtatious correspondence from Germany. He even took on the thankless task of correcting some of her sonnets. Riffaterre claims that the poem cannot be understood without knowledge of its 'hypogram', two images drawn from 'La Chevelure' and 'Harmonie du soir', the first linking by synaesthesia the olfactory and the visual in the perfume, hair, handkerchief image, and the second fusing the religious and the erotic in

'Variations sur la mort'; variations differing from one another, no doubt, but presumably also variations on the treatment of death by Gautier, Baudelaire, and others. And so with Paris and with the treatment of spleen. 'Litanies de mon triste cœur' is a poem modelled on Baudelaire's 'Les litanies de Satan', and, interestingly, two Baudelairean motifs, 'le bourdon' and 'l'horloge', were late additions. The Baudelairean vocabulary in the poem 'Spleens exceptionnels' ('fêtes', 'remords', 'cuisses', 'géantes') may well, like its title, be intended to evoke a smile of recognition.

The same perhaps applies to 'Sonnet de printemps', where Laforgue's lines recall the delicate fantasy of Gautier's well-known poem, 'Premier Sourire du printemps':[41]

Avril met aux buissons leurs robes de printemps
Et brode aux boutons d'or de fines collerettes.

Gautier's verse and all poetry of this kind appear to be gently mocked when Laforgue turns the sonnet towards death instead of the hope and love suggested by spring:

Hélas! j'attends toujours toujours l'heure sereine,
Où pour la grande nuit dans un coffre de chêne,
Le Destin ce farceur voudra bien m'emballer.

the phrase 'au concert rythmé des encensoirs'. The reader is frustrated and challenged, he argues, by the 'riddle' with which he has been presented and which he must solve before the poem can begin to be understood. In my view, this poem in Laforgue's early style would not have seemed obscure to his contemporaries—who were as much steeped in Baudelaire as he was, and imitated him just as much—nor does it seem so to the modern reader, since these are two of Baudelaire's most famous poems, whose images have become common currency. The poem seems to me to represent an example of the rather obvious pastiche common in young poets who are learning their trade. What interest it contains lies in the variation of the familiar Baudelairean climate that Laforgue introduces, drawing the poem slightly towards one that he will make his own. Riffaterre's insistence on obscurity depends in part on his reading of 'sa chevelure' as 'its hair' in the lines: 'Un relent énervant expire d'un mouchoir, | Et promène sur mes lèvres sa chevelure.' But Laforgue would probably (in my view) have intended the more natural, if less strictly grammatical, reading of 'her hair' (the hair belonging to the owner of the 'corps soyeux que j'adore'), which could easily have been introduced in a salon situation by vocal emphasis. This seems to me likely, because the 'hair' of a perfume emerging from a handkerchief is an image quite foreign to the poetry of the 21-year-old Laforgue—or, for that matter, to Laforgue at any period.

[41] T. Gautier, *Émaux et camées*: 'Pour les petites pâquerettes | Sournoisement lorsque tout dort, | Il repasse des collerettes | Et cisèle des boutons d'or.'

This is not a remarkable poem, but the method is one that will become typical of Laforgue, as we can see, for example, from the final stanza of the last of the *Derniers Vers*, which echoes Baudelaire:[42]

> O Nature, donne-moi la force et le courage
> De me croire en âge.[43]

These examples show how the model, functioning as a recognizable 'given', inspires a variation from it. Many of Baudelaire's themes and motifs, as well as those of other poets, are useful to Laforgue in a more general sense. They provide him with a starting-point. Observations and nuances of feeling are superimposed on to the pre-existing canvas. This happens with Paris when Laforgue forgets the *idée reçue* of the suffering poor and the complacent bourgeois. It is found in the development of details such as the use made of the familiar motif of the barrel-organ. Laforgue tries out variations on it: 'quelque orgue pleurard',[44] 'vieil orgue éreinté',[45] 'pauvre orgue monotone',[46] and in one line he looks forward to the two *complaintes* that he will write directly about it: 'un orgue inconsolable ululant ses complaintes'.[47] It is most striking in the image of the sunset, which is seen in Baudelairean splendour ('Le soleil écarlate | Là-bas descend tranquille, en une gloire d'or'[48]) but which can also already be 'douloureux',[49] 'jaune et poussif',[50] in a décor that is 'poitrinaire et macabre',[51] prefiguring the sun of the first poem of the *Derniers Vers*, which is compared to the spittle outside a drinking-dive. Autumn is already turned away from romantic splendour to winter. Twilight, mist, wind, and rain are still given solemn and explicit treatment, but they begin to take on the personal resonances of desolation that they will keep throughout Laforgue's work. Occasionally we find original images that are to become Laforguian leitmotifs. The piano-playing in well-to-do suburbs is an example. Twice given a line in this volume,[52] it will become central in the 'Complainte des pianos qu'on entend dans les quartiers aisés', and will still be present as a theme in Laforgue's last poems.

The serpentine paths towards original self-expression are well illustrated in this early work. One example is the way in which the conventional theme of fear of death, seen in Hamlet-like addresses to a skull

[42] Charles Baudelaire, 'Un voyage à Cythère' (*Les Fleurs du mal*): 'Ah! Seigneur! donnez-moi la force et le courage | De contempler mon cœur et mon corps sans dégoût!'

[43] Jules Laforgue, *Derniers Vers* XII.

[44] 'Les Boulevards'.

[45] 'Hue, carcan!'.

[46] 'Incurablement'.

[47] 'Dans la rue'.

[48] 'Le Sphinx'.

[49] 'Couchant d'hiver'.

[50] 'Pâle Soleil d'hiver'.

[51] 'Couchant d'hiver'.

[52] 'Triste, triste', 'Les Après-midi d'automne'.

and a preoccupation with the macabre that no doubt owes much to Gautier and Baudelaire, leads directly to what is to become the very Laforguian theme of insecurity and fear of life:

> Seul, sans amour, sans gloire! et la peur de la mort!
> Et la peur de la vie, aussi! Suis-je assez fort?[53]

Conventional expressions of romantic longing, such as are found in 'Sieste éternelle', prepare the way for the very personal theme of sexual frustration, stated in all simplicity:

> Je puis mourir demain et je n'ai pas aimé.
> Mes lèvres n'ont jamais touché lèvres de femme . . .[54]

Laforgue's attendance at the young poets' Club des Hydropathes may have led him to be more unreserved sexually than was usual at the time in serious verse. Whether he was or was not the first French writer to use the word 'clitoris'[55] in verse, this reference, as well as that to the sexual act as 'un accès de trois minutes',[56] point forward to his later use of sexually explicit references and neologisms. The conventional opposition of rich and poor takes him to some precise and disturbing observations of the poor not merely as objects of compassion, but for their ability to convey life's imperfection and ugliness, for life is already viewed from an aesthetic angle. 'Puis-je me plaindre de n'être pas heureux', he asks, 'Quand il est des femmes à jamais laides?'[57] In *Derniers Vers* VI he was to denounce the cruelty of life in the same somewhat unexpected terms:

> Nous ne serons jamais plus cruels que la vie,
> Qui fait qu'il est des animaux injustement rossés,
> Et des femmes à jamais laides. . .

Baudelaire's use of maternal imagery for sexual love may have given him the idea of using it in a more straightforward manner. We find examples of an imitative Baudelairean usage,[58] but we also have protestations of simple maternal need drawn from personal experience. Laforgue describes both a nostalgia for lost childhood and grief at his mother's death in the barest confessional tone:

> Je voudrais être enfant, avoir ma mère encor
> ('Les Après-midi d'automne')

[53] 'Les Après-midi d'automne'.
[54] 'Pour le livre d'amour'.
[55] 'Guitare'.
[56] 'Pour le livre d'amour'.
[57] 'Fragment'.
[58] 'Excuse mélancolique'.

Je vis!—Pourtant, ma mère est là-bas au tombeau!
('Rabâchages')

The image of the bleeding heart that encompasses man and the universe and is central to the volume must owe much to Baudelaire's (and other poets') personification of the setting sun:

Le soleil s'est noyé dans son sang qui se fige.[59]

Laforgue sees his own heart mirrored in the daily bleeding to death of the sun, principle of life and sexuality. 'Je veux saigner pour tout, saigner toujours, encor',[60] he writes, and sees himself as the heart of the universe: 'Le cœur de l'univers est né: c'est moi qui l'ai, | Je suis le cœur de Tout!'[61] The objects of the real world cast back reflections of suffering, the great rose-window of Notre-Dame, or the cathedral organ that is a 'cœur trop plein d'amour'.[62] But there is a satisfaction to be gained from this fluid image of a constant seeping of the self into an environment that mirrors it. In Baudelaire it is the 'vaporisation'[63] linked with artistic creation and described in 'Confiteor de l'artiste' as a diffusion of the self into nature, 'car dans la grandeur de la rêverie le *moi* se perd vite'. Laforgue writes: 'Mon être se dissout',[64] 'Et je rêve qu'ainsi je m'éparpille aux cieux',[65] and clearly in this process he must have felt that he was giving in to Hartmann's unconscious, which 'échappe absolument au regard de la conscience'.[66]

Loss of a sense of personal identity leads to what is perhaps the most forward-looking element in this volume, the questioning of that identity. In 'Suis-je?' Laforgue asks with desperate insistence the questions that he will address with wit and urbanity later, in 'Ballade' for example:

Mais suis-je seulement, insensé!
Quel vertige!

. . . .

Ce corps renouvelé chaque jour est-il mien?
Je vois un tourbillon, incessant va et vient
D'atomes éternels, oublieux, anonymes,

. . . .

Oui, ces poumons, ce cœur, cette substance grise
Est-ce Moi?
('Suis-je?')

[59] Baudelaire, 'Harmonie du soir', *Les Fleurs du mal.* [60] 'Désolation'.
[61] Ibid. [62] 'Incurablement'.
[63] Baudelaire, 'Mon cœur mis à nu', *Œuvres complètes*, i, ed. C. Pichois (Paris, 1975), 676. [64] L'Oubli'. [65] 'A Saint-Cloud'. [66] Hartmann, *Philosophie*, 321.

Oyez, au physique comme au moral,
Ne suis qu'une colonie de cellules
De raccroc; et ce sieur que j'intitule
Moi, n'est, dit-on, qu'un polypier fatal!
De mon cœur un tel, à ma chair védique,
Comme de mes orteils à mes cheveux,
Va-et-vient de cellules sans aveu,
Rien de bien solvable et rien d'authentique.
Quand j'organise une descente en Moi,
J'en conviens, je trouve là, attablée,
Une société un peu bien mêlée,
Et que je n'ai point vue à mes octrois.
('Ballade')

There are lighter touches even in the poems intended for this volume. This is not surprising, because Laforgue had written a comedy in verse, *Tessa*, in 1877, and a series of occasional poems around 1879, some of which were published in the Toulousian literary review, *La Guêpe*. Their clear intention was to surprise, shock, and amuse. They are occasionally subtle, as in the ironic last line of 'Intérieur', a poem describing a woman sitting sewing in a room, dreaming of a richer world outside, a theme that Laforgue will develop later in images of imprisoned 'jeunes filles'.

Elle écoute le vent, rève parfois, s'arrête,
Tire la longue aiguille et s'en gratte la tête;
Puis reprend aussitôt, avec son air songeur.
Et moi j'intitulai ma pièce: *Intérieur*.

Even in the poems destined for *Le Sanglot* we find the occasional ironic self-portrait:

Par l'azur tendre et fin tournent les hirondelles
Dont je traduis pour moi les mille petits cris.[67]

or

Si je rime au Néant c'est pour jouer mon rôle.[68]

'Avant d'avoir vécu dégoûté de la vie'[69] is unsparingly accurate, while 'Je tords mon cœur pour qu'il s'égoutte en rimes d'or'[70] is reminiscent of the later description of his poetry as 'perles curieusement taillées'.[71]

67 'Crépuscule de dimanche d'été.' 68 'Éponge définitivement pourrie'.
69 'Lassitude'. 70 'Veillée d'avril'. 71 *AH* 763.

There are also examples of a 'langue fouillée', for example, the *transposition d'art* of the following lines:

> A travers le lacis des branches dépouillées
> Dont l'eau-forte sabrait le ciel bleu-clair et froid.[72]

A light self-mockery is seen in the two stanzas of 'Nocturne' that neatly oppose the morose poet to three charming girls, blithely indifferent to the sufferings of the universe:

> Je songe au vieux Soleil un jour agonisant,
> Je halète, j'ai peur, pressant du doigt ma tempe,
>
> En face, pourtant, trois jeunes filles causant,
> Brodent à la clarté paisible de la lampe.

Marginal notes on the manuscripts are often humorous also: the writing of 'immense' with three *m*'s ('i*mmm*ense') at the side of the poem that contains the line, 'Comme d'une nausée immense de la vie',[73] for example. Presumably the comment, 'Pour faire pendant au lac de Lamartine', above the poem 'Paroles d'un époux inconsolable' was not serious.[74] Exactly what tone we should attribute to the following marginal comment on his sonnet-writing is not clear, but it sounds like irritation at the sameness of his productions: 'Toujours le même moule mes sonnets: dans les deux quatrains le décor, puis l'idée dans les deux tercets et le vers final.'[75] If it is critical, we can perhaps read the single line that appears on two of the three manuscripts of the sonnet 'Berceuse', after the poem, as an attempt to break the mould: a challenge to the sonnet form as well as to the acceptance that all is vanity, which is the main theme of the lullaby. The alexandrine 'Mais moi je ne veux pas m'en aller sans savoir', then, rather than being a line 'sans rapport, semble-t-il',[76] as the editors of the complete works suggest, would play a role similar to the fifteenth line that Éluard added to his sonnet 'Le Jeu de construction';[77] that is, reverse the earlier direction of the poem, establish an area of dialogue, and end the poem on a conflicting note. All these characteristics are found in the *Derniers Vers*, where final stanzas frequently have this deliberately destabilizing effect.

On the whole, however, the poems intended for this volume are explicit and solemn because this was Laforgue's intention. The hints of

72 'Couchant d'hiver'. 73 'Bouffée de printemps'.

74 The fact that Laforgue's father planned a study of Lamartine cannot be insignificant. 75 By 'Triste, triste'. 76 *AH* 384.

77 P. Éluard, 'Mourir de ne pas mourir', *Capitale de la douleur*.

future directions towards an impressionistic rather than a philosophical style, towards irony and the 'modern', remain hints. Much was sacrificed or distorted to produce the desired single-tone 'note aiguë'. However, while writing in a manner that was so fundamentally opposed to the one he was later to adopt, Laforgue used techniques that persisted, and which had their origins in contemporary writings. First, the oratorical tone of the verse, dependent on Hugo, Vigny, Leconte de Lisle, and others. Laforgue was never to abandon the dramatic forms that are already present here. He may shift addressees from sun and universe to woman or life, but he continued to set up dramatic confrontations and to present them as such to the reader. An initial dualism in the poet's response to the world, no doubt to some extent influenced by Baudelaire, is reflected already in the vocabulary in this volume, a minimalist vocabulary to describe the narrator, with much emphasis on the poet as child ('Je serais votre enfant',[78] un pauvre enfant vidé',[79] 'Pauvre petit cœur sur la main',[80] 'J'entends mon cœur qui bat, | C'est maman qui m'appelle'[81]), and a hyperbolic vocabulary to describe the vastness of the universe and the intensity of the poet's sufferings. When the 'grandes angoisses métaphysiques' have become merely 'chagrins domestiques',[82] the hyperbolic tone that is central to this volume, and which Laforgue explored fully in his attempt to find a language that did justice to his suffering, can quite naturally serve his irony, conveying simultaneously by its overstatement both the suffering and a narrative distance from it.

Hyperbole was accompanied by exclamations and rhetorical questions, which are typical also of Laforgue's latest poems. Repetition, which is used here for dramatic emphasis, will later be used to express the disjointed conversation with the self, the 'balbutiement' of the '"tête-à-tête" avec soi'.[83] Enumeration is used here for breadth and grandeur ('Gloire, Douleur! Douleur! Encore!'[84]). Later it will become recapitulatory and low-key, often functioning in the manner of a theatrical aside, and Laforgue will use it to create a jarring sense of the different levels of the mind. The refrain is another technique that Laforgue does not abandon, as is the use of the one-line stanza. He already employs to an unusual degree a *langage parlé*,[85] deriving from the

78 'Excuse mélancolique'.
79 'Les Spleens exceptionnels'.
80 'Pauvre petit cœur sur la main'.
81 'La Chanson du petit hypertrophique'.
82 'Complainte d'une convalescence en mai'.
83 *LA* 54.
84 'Devant la grande rosace en vitrail, à Notre-Dame de Paris'.
85 'D'autres ronflent ou font l'amour, *et cætera*' ('Rabâchages'); '*Bah*! Sortons' ('Spleen'); 'Ah! triste, triste, *va*, triste' ('Soleil couchant de juin').

dramatic or intimate vocabulary of Hugo or Baudelaire, but also reflecting his writing for *La Guêpe* and his frequentation of the Club des Hydropathes, where poems were read aloud. Laforgue's next volume, *Les Complaintes,* deliberately sets out to cultivate 'les refrains des rues', while in the *Derniers Vers* the colloquial is vital as a springboard for the lyrical.

It has been suggested that the usual path of a literary influence is from imitation, through rejection, to assimilation. As we have seen, assimilation is present in Laforgue's case, but in a somewhat perverse fashion. In the textual sense that we have just been considering it seems to go unrecognized by its author, presumably because the borrowed elements are used for purposes that are quite alien to those of precursors, who seem most important to the young author as models to be rejected in the search for the original and the novel. Perhaps in this agonistic[86] sense Rimbaud's debt to his unacknowledged predecessors was also greater than he realized.

[86] Harold Bloom develops this approach in *Agon* (New York and Oxford, 1982) and many of his other critical writings.

2

Les Complaintes: 'Les refrains des rues'

THE IDEA OF THE *COMPLAINTE*

'LES vers pompeux sont embêtants',[1] wrote André Gill, and by 1882 Laforgue agreed with him. Having distanced himself from the poets whom he had at first imitated, he found for his next volume a quite different model. It was surprisingly remote: the plaintive and burlesque *complainte* of the sixteenth century, but Laforgue coupled this with later folk-songs, down to the doggerel jingles of his day, a combination of traditional popular proverb and 'refrains des rues'[2] of the contemporary world. The model was useful to him chiefly in two ways. Since the *complainte* was a popular genre, intended to be spoken or sung, Laforgue was released from traditional elevated verse, and solemn self-absorption became technically impossible. The tone adopted by the *complainte*, although it was an antiquated form, established a 'modern' down-to-earth familiarity and realism, which forced him to attempt to implant the immediacy of the oral into the written. Secondly, his *complainte* was a 'rewriting', a variant on an earlier text. He was forced also to deviate from his model: that is, to write indirectly and ironically. Laforgue's *parlando* style and his irony both develop therefore from this inspired move.

Variations on a recognizable model create a variety of effects. They invite us to re-examine the human 'truth' behind the familiar but discarded model: they offer us a new view; they offer it in such a way that it is clear that it also is relative and likely to be superseded. If the new view is itself undercut by irony, we cannot escape the realization that we inhabit a world that offers nothing more substantial than subjective and shifting impressions. Laforgue's talent for responsiveness, his floating sensibility, his wit and intellectual acuteness were all called into play by

1 A. Gill, *La Muse à Bibi* (Paris, 1880).

2 J.-L. Debauve, *Laforgue en son temps* (Neuchâtel, 1972), 113. 'Je n'aurais jamais rêvé un article aussi intelligent. Le passage où il dit que j'ai recueilli "les refrains des rues, des bois, de l'alcôve, de l'église, de la causerie bourgeoise, des grands discours, du peuple et de la solitude, de la terre et de l'air" est d'une surprenante divination.'

this approach, without his having the possibility of making a direct comment on the world. He could now cultivate the enigmatic and the elusive, the qualities that had no place in philosophical verse, but which were central to his artistic personality and, by now, to his credo.

His new aesthetic emphasized the perfume rather than the flower—an analogy he drew directly from Bourget.[3] 'Je rêve de la poésie qui ne dise rien', he wrote, 'mais soit des bouts de rêverie sans suite.'[4] He also spoke, as we have seen, of 'de la psychologie dans une forme de rêve', and of 'd'inextricables symphonies avec une phrase (un sujet) mélodique, dont le dessin reparaît de temps en temps'.[5] In the copy of the *Complaintes* that he sent to his sister Marie, conscious that she would find the poems strange and probably incomprehensible, he insisted that he had not changed at heart: 'Il n'a pas changé, ce cœur. Il est toujours aussi gros. Il est devenu un peu plus littéraire, voilà tout.' By many of the poems he noted particular dates and events to show that they sprang from his personal life.[6] It was the distancing from these origins that was now aesthetically important, however, because, in Laforgue's words, 'Une poésie ne doit pas être une description exacte, (comme une page de roman), mais noyée de rêve.'[7] The subjects could be mundane—the barrel-organ, for example—their poetic value would depend on the poet's inner world or his 'fantaisie'. So, simultaneously, in Baudelaire's famous definition, the reader is offered object and subject: 'le monde extérieur à l'artiste et l'artiste lui-même'.[8] The genre of the Laforguian *complainte* depends on this fusion, unexpectedly creating an interior monologue from the most trivial elements of the external world. Not the least of the effects frequently created in the *Complaintes* is that of the pathos of objects: pianos and barrel-organs, an abandoned villa or an empty casino, hospital beds or old photographs; less obviously, but equally poignantly, such things as a Sèvres vase illustrating a pastoral

[3] *AH* 822. [4] Ibid. 792. [5] Ibid. 757.

[6] Thus 'Complainte-litanies de mon sacré-cœur' and 'Complainte des condoléances au soleil' have 1881 and 'rue Berthollet' written beside them. (The dates may sometimes refer to the experience that inspired the poem rather than to its final composition.) Other dates are 'Complainte des pubertés difficiles': Sept. 1882; 'Complainte des grands pins': Oct. 1882; 'Complainte d'une convalescence en mai': 1884; 'Complainte-variations sur le mot "falot-falotte"': 1884, and also the place 'Chevreuse'. Since 'Complainte du pauvre jeune homme' and 'Complainte de l'époux outragé' have simply 'Chevreuse', 1884 is probably understood for them as well. In Laforgue's *Agenda* for 1883 the entry for 30 July refers to 40 finished *complaintes* (*AH* 894). The final number was, of course, 50. [7] *AH* 821.

[8] Charles Baudelaire, *L'Art philosophique*, *Œuvres complètes*, ii, ed. C. Pichois (Paris, 1976), 598.

idyll, or a lopsided window-blind on which hangs a forgotten pair of gaiters.

In 'Plainte d'automne', a prose poem that, as we have seen, Laforgue admired,[9] Mallarmé had asked why the vulgar music of the barrel-organ, which can, of course, be seen as the musical equivalent of the folk-song, should have the power to move him deeply: 'Maintenant qu'il murmurait un air joyeusement vulgaire et qui mit la gaîté au cœur des faubourgs, un air suranné, banal: d'où vient que sa ritournelle m'allait à l'âme et me faisait pleurer comme une ballade romantique?' What has been described as 'une esthétique de la platitude',[10] is based on the paradox of the banality that gives rise to deep emotion, and fittingly, it owed its origin to an event similar to that referred to by Mallarmé.

It was the celebration of the inauguration of the lion of Belfort, the statue in the place Denfert Rochereau, on 20 September 1880, and Laforgue recorded it vividly in a fragment that reveals the triviality of the occasion, as well as his own sense of alienated desolation.

—Fête de nuit. Inaugurat. du lion de Belfort—pauvre—triste—temps triste. place d'Enfer, Observatoire, fête foraine. Des chevaux [de] bois tournant. des balançoires. des marchands de ferrailles. des faiseurs de caramels. des somnambules, des tourniquets où des étudiants ont gagné un vase de nuit au fond duquel un œil en émail peint regardait. Un cirque avec des toiles grossièrement peintes éclairées par des quinquets fumeux et fétides. deux femmes en maillot fané se promenant sur les planches, gueulant. Des musiciens faisant rage dans des cuivres bosselés dominés par la grosse caisse. boum! boum! Un paillasse avec un large pantalon, montant jusqu'au cou et serrant les chevilles, au dos une horloge brodée, perruque d'étoupe rouge, chapeau pointu blanc, masque de farine qui se plissait, se ridait quand il se pâmait sans conviction (ce monsieur, ce frère a ses soucis comme vous et moi—Drôle!)

Une noce entière occupait un manège de chevaux de bois. la mariée en jupon sali dans tous les gargots graisseux de l'arrondissement, se disputait avec le loueur, lui mettait ses deux poings sous le nez. Le marié bêtement s'esclaffait. Une femme de la noce vomissait des flaques de vin, où un chien lappait. Une autre lui tapait maternellement dans le dos pour exciter, faciliter; bougonnant—il n'y avait pas de bon sens après avoir bu et mangé toute la journée à aller tourner sur des chevaux de bois.

Des ménages d'ivrognes. Un souteneur faisant sortir une bande de filles dont l'une adorable et triste avait un bleu sous l'œil, elles buvaient du vin—odeurs de quinquets, glapissement des montreurs, mélancolie des orgues jouant des airs de

[9] *AH* 693.

[10] J.-P. Richard, *Pages paysages: Microlectures*, ii (Paris 1984), 44.

carrefours d'automne. en haut les étoiles vierges et éternelles—Drôle de planète![11]

In this passage Laforgue observes the human pathos of the failed festivity. Dissonances prevail: the crude student humour, the hints of dingy poverty, the clowns' unconvincing gestures, the vulgar arguments of bride and groom, the bride's soiled dress, the pretty young prostitute with a black eye. An outsider, Laforgue could have written with Charles Cros: 'Moi je vis la vie à côté, | Pleurant alors que c'est la fête.'[12] He was to relive even these memories 'à côté', as he revived them for the *Complaintes* amid the luxury of the German court, which, while it suited his innate yearning for elegance, contributed a further sense of exile. Already in these 1880 notes he registered an ironic distance from the scene, with his concluding 'Drôle de planète'. The *complainte* emerged specifically from the singing of the two women, an unaccompanied singing that Laforgue described unromantically as 'gueulant'. That and other refrains, of course: 'Pour goûter cette chose', he wrote later to Kahn, 'il faudrait chanter les refrains sur un air de cor de chasse que j'ai entendu dans mon enfance en province.'[13] But it took time for him to see that he must in a sense *be* these women, and that he must impersonate not only them, but the many voices at the festivity. The truth that the 'sincere' tone was inevitably insincere, that only oblique methods could hope to capture the actual complexity of his response to life had still to be fully assimilated. He continued, at first, simply to record his personal lament:

> Oh! la vie est trop triste, incurablement triste!
> Aux fêtes d'ici-bas j'ai toujours sangloté.
> ('Soir de carnaval')

or to describe such scenes objectively:

> J'errais par la banlieue en fête, un soir d'été.
> Et, triste d'avoir vu cette femelle enceinte
> Glapissant aux quinquets devant sa toile peinte,
> Près des chevaux de bois je m'étais arrêté.
> ('Hue, carcan!')

Two early *complaintes*, 'Complainte de l'organiste de Nice' and 'La Chanson du petit hypertrophique', which Laforgue rejected for his volume, prefigure his later style. In both of them the main character is a

[11] *AH* 650-51.
[12] C. Cros, 'Sonnet', *Le Collier de griffes* (Paris, 1908).
[13] *LA* 56-7.

double, on whom the poet projects his suffering. He probably rejected them because of the strong emotional element that they both contain, allowing into his volume rewritings of early poems with a more obviously satirical slant,[14] but this now seems a mistake. Both are subtle poems that show how cliché can be rejuvenated to serve effects of pathos. In the first Laforgue describes the artist's exaggerated sorrow at the imagined death of a girl he does not know, and the final irony is directed at the poet for trying to compose a poem from so 'fictional' a subject. The idea is one that Laforgue did not abandon: he developed it in a later *complainte*[15] and in the eleventh of the *Derniers Vers*. The irony of 'La Chanson du petit hypertrophique' is of a similar kind. There is no attempt to mock the sufferings of the child narrator, but childish slang and metrical abbreviations ridicule the kind of poetry that might be expected to result from such a theme.

The simplest form of *complainte* used by Laforgue is that based on a popular song.[16] The well-known words and tune form a background to the revised version. The method—if we consider the 'Complainte du pauvre jeune homme', based on 'Quand le bonhomm' revint du bois'—takes a tragic narrative, here the suicide of a young man whose wife has abandoned him, and treats it with the anonymity and detachment of a newspaper item. As with Mallarmé's barrel-organ, the insistently cheerful rhythm drives home the pathos of the banal but tragic plot. The simplicity is deceptive, and must be so if the poem is not to be merely trivial itself. The young man is a persona of the poet, sensitive, lonely, given to ennui. His wife's desertion causes the tragedy, but the scene has already been well prepared. His is a 'belle âme', ill at ease in the modern world. While nostalgia for the nobility of soul of former happier times pervades the poem, it is simultaneously mocked: a *belle âme* is not merely something not 'found' in modern times: it is something, apparently man-made, that is no longer 'produced'. Laforgue's 'rewritings' complicate and fuse worlds, as do the neologisms that he invented most successfully at this period, and which he described as 'cet accouplement de mots qui n'ont qu'une harmonie de rêve mais font

[14] 'Complainte-litanies de mon sacré-cœur', 'Complainte des condoléances au soleil'.

[15] 'Complainte de la bonne défunte'.

[16] Henri Davenson (*Le Livre des chansons* (Neuchâtel, 1946), 17) distinguishes two categories of oral folk-song: 'les *rondes* ou chansons à danser que caractérise la présence d'un refrain', and 'les *complaintes* ou récits continus'. The categories were not, however, totally distinct even before Laforgue blurred them. Typical of the folk-song's attempt to imitate popular speech were two techniques that Laforgue frequently employs: the removal of the mute *e*, which is replaced by an apostrophe; and the inclusion of a *z* or a *t* to fill the hiatus between vowels.

dans la réalité des couples impossibles (et qui ont pour moi le charme insoluble, obsédant, entêtant des antinomies en métaphysique . . .)'.[17] In the 'Complainte de l'époux outragé', the rewriting of the little tale of adultery that is the subject of the song 'Qu'allais-tu faire à la fontaine?', Laforgue mixes religion and the erotic, parodying the solemn treatment often given to this combination in contemporary verse. In the 'Complainte de Lord Pierrot' the folk-song 'Au clair de la lune', a commonplace rhyme about the ruses of physical seduction, is the prelude to the most melancholy, idealistic, and rambling of *divagations* on the subject of sexual desire. 'Mon ami Pierrot' becomes the enigmatic 'Lord Pierrot', half English reticence, half *commedia dell'arte* mask. It is his 'cervelle' not his 'chandelle' that is dead, and the shared assonance ironically drives home the distance between the worlds that the two versions inhabit. As the later *moralité* 'Persée et Andromède' moves from one legend, that of Perseus and Andromeda, to a second, that of Beauty and the Beast, so here one folk-song, 'Au clair de la lune', gives way to another, 'Il pleut, il pleut bergère', appropriate to the poem's final plunge into melancholy.

In the 'Complainte de cette bonne lune' the folk-song 'Sur le pont d'Avignon' serves to introduce a celestial dance in 'l'giron du Patron' (God), in which the moon, a Cinderella-figure among the dazzling stars, far from finding her Prince Charming, rejects all invitations to the dance, because of her concern for her poor sister, the earth. So, instead of a picture in which 'all's right with the world'—and with heaven, too—which the original owed to the magical and religious associations of ritual dances on bridges, we are offered a humorous debunking of all such 'certainties', as the dialogue between the two parties degenerates into a vulgar brawl between celestial bodies. 'Est-ce assez idiot?',[18] Laforgue wrote to Kahn when he sent him the preposterously inventive 'Complainte du fœtus de poète', and one might ask the same of this poem.

There are rewritings that are less frivolous but equally transform their 'model'. Goethe's poem 'Es war ein König in Thule', the model for the 'Complainte du roi de Thulé', is a lyric celebrating marital fidelity, symbolized by the golden goblet bequeathed to the king by his wife on her death, which he hurls into the waves as his own death

[17] *LA* 64–5. Some examples are: *éternullité*, *violupté*, *madréporiser*, *feu-d'artificer*, *instincts-levants*, *sexciproque*, *voluptante*, *soleil-crevant*, *crucifiger*, *sangsuelles*. Laforgue wearied of them, and tried unsuccessfully to change 'sangsuelles' ('Complainte des voix sous le figuier boudhique') to 'sensuelles': Debauve, *Laforgue*, 150.

[18] *LA* 55.

approaches. Against this uplifting example, Laforgue sets one that he places even higher. His king substitutes renunciation for fidelity. Instead of the devoted couple, we have a solitary figure, described in a cosmic landscape of great splendour, descending to mysterious polar regions to console a dying sun. In the final stanza the image of the absolute, to which the 'real' is sacrificed, is offered to young lovers as a more elevated and inspiring goal than theirs—a vision of 'amour pur'. Goethe's romantic depiction of conjugal love in the face of age and death lingers behind a poem that offers a more modern and ambiguous image, the power of which depends on the beauty of exalted negation:

> Braves amants! aux nuits de lait,
>   Tournez vos clés!
> Une ombre, d'amour pur transie,
> Viendrait vous gémir cette scie:
> 'Il était un roi de Thulé
>   Immaculé. . .'

This poem, like the 'Complainte de Lord Pierrot', might well be described as 'de la psychologie dans une forme de rêve', but poems where the model is woven into the text, itself forming part of the dream, best fit this description. The 'Complainte des pianos qu'on entend dans les quartiers aisés' is an example. Here the *complainte*, 'Tu t'en vas et tu nous laisses | Tu nous laiss's et tu t'en vas', is introduced as the words that accompany the piano notes that the narrator, a solitary *flâneur*, hears as he wanders through the streets in the town's well-to-do suburbs. The poem is composed of four regularly repeated strains, each set in counterpoint to the others. Differing line- and stanza-lengths make the pattern perfectly clear: two belong essentially to the narrator, and are followed by two that represent the thoughts of the girls whom he imagines playing the piano in the seclusion of their homes or convent schools. The poem moves from present to future as the narrator envisages the loveless *mariages de convenance* for which the girls are destined in place of the romantic love to which they aspire and which is the subject of the *ritournelles* they practise on the piano. The whole poem—both the image of the narrator, similarly lost in an inner world that aspires to the romantic, and the image of the present desert and future emotional death reserved for the girls—has sprung from the overheard piano tune, itself inseparable from the words of a popular song. Two lines from the song, repeated five times, are subtly modulated by the lines that complete the quatrain, in each case offering a glimpse of a different aspect of the girls' lives, and moving in an imaginative progression that leads to the final brutal glimpse of a future, the reverse of their aspirations, from which they will look back to their bored piano-playing days with longing:

'Tu t'en vas et tu nous laisses,
Tu nous laiss's et tu t'en vas.
Que ne suis-je morte à la messe!
O mois, ô linges, ô repas!'

The *complainte* is not impenetrable, as has been suggested, but it is built on an elliptical technique that asks the reader to juxtapose and reconstruct its contrasting and mobile elements. The elements are themselves elliptical, and this is appropriate since the barely conscious world towards which the poem is directed speaks most convincingly in brief, intense ejaculations. The poem represents a journey inwards from the deceptively urbane and leisurely opening stanza, with its undercurrents of obsession and its reference to nerves where the heart would be more natural:

Menez l'âme que les Lettres ont bien nourrie,
Les pianos, les pianos, dans les quartiers aisés!
Premiers soirs, sans pardessus, chaste flânerie,
Aux complaintes des nerfs incompris ou brisés.

Laforgue's volume of *Complaintes* was found incomprehensible by most reviewers, and this despite the fact that the 'modern' and the burlesque were fashionable at the time, with the supposed Adoré Floupette's *Les Déliquescences*,[19] which appeared in May 1885, being an example of a more frivolous excursion into parody. The tradition by now included Charles Cros's *Le Coffret de Santal* (1873), Corbière's *Les Amours jaunes* (1873), and Richepin's *La Chanson des gueux* (1876). But the *Complaintes* were just too baffling, too 'modern'—'de l'ultra-moderne', as one perceptive reviewer put it: 'Si vous aimez la vraie modernité, pas celle d'hier ni d'aujourd'hui, mais celle de demain, je vous conseillerai . . . *Les Complaintes* de M. Jules Laforgue. C'est de l'ultra-moderne.'[20] The majority of reviewers felt that they were being fooled. 'Si ça continue', one wrote, 'il suffira dans six ans . . . d'écrire comme un Javanais: pour être un poète de génie.'[21] Laforgue wrote one review himself, made emendations to another, written by Charles

19 The authors were Henri Beauclair and Gabriel Vicaire.

20 Debauve, *Laforgue*, 210.

21 Ibid. 198.

Henry, and was amazed at the perspicacity of one critic, Léo d'Orfer.[22] His review stressed the remarkable range of the *Complaintes*, the aspect that we shall consider next. But it also described what has been the present subject: how 'des lambeaux de refrains populaires, des demi-couplets de vieilles romances criaillées dans les cours, . . . toutes les chansons et chansonnettes des rues, des bois, de l'alcôve, de l'église, de la causerie bourgeoise, des grands discours, du peuple et de la solitude' could be accommodated to 'la sauce de la complainte'.[23] 'Pour ma part', d'Orfer wrote, 'j'avais rêvé un genre de poèmes où tous les prosaïsmes et les vulgarités de la vie réelle trouveraient place à côté d'envolées superbes, les uns étant l'intelligence des autres.'[24] This combination was what he had found in Laforgue's *Complaintes*, and it is one that the twentieth century, schooled to an aesthetic that finds dissonance indispensable, is prepared to take seriously.

MULTIPLE VOICES

J.-P. Richard points out the variety and ambiguity to be found in even the titles of the *Complaintes*. Some are votive laments, pleas dedicated to an external power ('Complainte propitiatoire à l'Inconscient', 'Complainte à Notre-Dame des soirs'); some are laments whose theme is an aspect of the insufficiency of life ('Complainte sur certains ennuis', 'Complainte sur certains temps déplacés'); most use the genitive form (*de*, *du*, *des*), but, since they employ both subjective and objective genitives, even this does not confer unity on them. The reader is, as Richard suggests, disorientated:

> Il arrive pourtant que l'ambiguïté, toujours grammaticalement possible, de la préposition s'actualise peu ou prou dans l'énoncé de tel ou tel titre, et que le lecteur ne sache plus dès lors, au cœur d'un petit trouble signifiant, si l'être nommé dans la seconde partie de la séquence titre est celui qui porte la parole ou celui que la parole vise.[25]

As we have seen in the case of the 'Complainte des pianos', the poem cannot be read simply as the lament of the pianos. It is dependent on a range of agents, and illustrates a number of subjectivities: on the one hand, the notes of music, the popular song, the Catholic bourgeois setting; on the other, the 'psychologies' of the girls, the narrator, and the

[22] Ibid. 113. [23] Ibid. 205. [24] Ibid. [25] *Hidd. LA* 88.

poet. The result, in the laments of musical instruments, is a situation in which the discourse (again in Richard's words) 'tout à la fois s'adresse à eux [musical instruments], traite d'eux et constitue une transposition (phrasée) de ce qu'ils sont censés prononcer (directement) ou connoter (indirectement) sur le mode musical', and the final effect is 'un certain flou de l'énonciation qui répond, on le sait, à l'un des effets les plus vivement recherchés par la poétique laforguienne: la délocalisation du moi, la confusion discursive du sujet et de l'objet, le demi-naufrage (joué) du sens.'[26] This effect is further increased by the fact that the representation of a multiplicity of voices is a deliberate aim; voices that are merely overheard, that emerge from a void, mobile voices representing moments in a changing narrative, and whose mobility matches that of the equally variable moods and personae of the narrator. Laforgue had learnt from Hartmann, as Warren Ramsey points out, 'to think of the human individual as an aggregate, a sum of many individuals'.[27]

In this climate of ambiguity and uncertainty what order and consistency do we find? Y.-A. Favre argues that we have a 'livre' rather than an 'album';[28] that is, a highly structured whole, and there is evidence that this is the case with all Laforgue's volumes, with the exception of the *Derniers Vers*, about whose intended final structure we know almost nothing. Plans were not rigid, however. The original *Complaintes*, significantly called simply *Quelques Complaintes de la vie*,[29] first numbered twenty poems, then grew to forty and to fifty with the help of Laforgue's publisher Léon Vanier, 'Fabius Cunctator', as Laforgue named him.[30] The new poems were interpolated casually, and when Laforgue sent Vanier a fiftieth poem, he told him to place it 'n'importe où'[31] in the volume. He insisted on the placing of two key poems, however: the opening 'Préludes autobiographiques', which was to be 'answered' by the late 'Complainte du Sage de Paris': 'Cette préface explique la dernière et longue litanie qui ferme le volume.'[32] But, with characteristic flexibility, he added two short poems to this litany, closing his volume in a more oblique and light-hearted fashion. One is the 'Complainte des complaintes', which serves as an apology for, and defence of, the volume: the other, the 'Complainte-épitaphe' uses

26 Ibid.

27 Warren Ramsey, *Jules Laforgue and the Ironic Inheritance* (New York, 1953), 118.

28 *Hidd. LA* 91-103.

29 *AH* 821.

30 *OC*, v. 125: 'On décernera à Vanier le titre de Fabius Cunctator—parce que, à supposer qu'un poète lui confie un manuscrit payé, que ce poète s'appelle *Cunct* et qu'il compte avoir son livre au jour convenu, on pourra dire de lui: Cunct a tort de compter, etc.'

31 *LA* 88.

32 Ibid. 79.

disyllabic quatrains and tercets in a parody of the sonnet form. Pointing forward to future exercises in ellipticism, such as 'Avant-dernier Mot',[33] and taking a leaf from various contemporary *fumiste* productions,[34] it sums up the themes and the method of the volume, the latter by reference to other art-forms, in a supreme distancing act:

La Femme,
Mon âme:
Ah! quels
Appels!

Pastels
Mortels,
Qu'on blâme
Mes gammes!

Un fou
S'avance,
Et danse.

Silence . . .
Lui, où?
Coucou.

'La Femme', 'mon âme' are, of course, seen through a dedication to the unconscious, which, deliberately blurred with an ascetic Buddhism, is the subject of the early votive poems. Even the 'Complainte-Placet *de* Faust fils' is, as Favre points out, a plea *to* nature.[35] (It is simultaneously a parody of Sully Prudhomme's sentimental poem 'Prière' and of Goethe's Faust's intellectual and guilt-ridden universe.) The volume represents a personal and moral journey from rejection and disillusionment to some kind of acceptance or resignation, a *schema* surprisingly parallel to Laforgue's plans for the *Sanglot*, a fact that might easily be overlooked because of the new volume's contrasting register and because love, which was largely absent from the *Sanglot* poems, has now become a central theme.

The second poem of the volume, 'Complainte propitiatoire à l'Inconscient', already sets up the technique of parallel discourses that we saw in the 'Complainte des pianos'. Here, more simply, the leading couplet takes the form of a prayer addressed to the unconscious, while the following quatrain develops an interior monologue in which 'la

[33] Discussed in Ch. 4, third section.

[34] An excellent anthology of contemporary *fumisme* is: Daniel Grojnowski and Bernard Sarrazin, *L'Esprit fumiste et les rires fin de siècle* (Paris, 1990).

[35] *Hidd. LA* 94.

Pensée', from which the unconscious should deliver man, is, all too obviously, failing to find a solution to his distress. Both discourses follow a linear progression: the couplets offer ordered parodic echoes of the Lord's Prayer ('Votre Nom', 'Volonté', 'quotidienne', 'Pardonnez-nous nos offenses', 'délivrez-nous'), while the quatrains move through the obvious range of possible human ideals: love as an absolute, the Christianity of the mystic, that of the missionary, the 'religion' of art. The intellect from which man is to be delivered is both a 'lèpre originelle' and an 'ivresse insensée', since it has created these false and impossible ideals. The advance in technique over such a *Sanglot* poem as 'Marche funèbre pour la mort de la terre', which also functions by ordered antithesis, is striking. In 'Marche funèbre' the refrain is unchanging and the monologue explicit. Here we inhabit immediately a puzzling, elliptical, and challenging world, with parody at its centre.

Fifteen of the *Complaintes* are composed on this model, which can, as we saw in the 'Complainte des pianos', become more intricate. In that poem each voice was subdivided. In other *complaintes* the dialogue is extended by a third voice, frequently that of a narrator, who either introduces the poem, encloses it in a first and final stanza, or develops a contrasting point of view in its conclusion. The two opposed but interrelating main voices usually contrast a general with a particular view. In the 'Complainte de l'orgue de barbarie', for example, the barrel-organ regurgitates the commonplaces of life in disabused and cynical five-syllable quatrains. These alternate with couplets—a stanza not commonly employed for the nerve-centre of a poem—expressing individual emotion: here that of a woman, and moving from ecstasy to fear and despair. These ordered snatches of an individual's life, which nevertheless illustrate the *idées reçues* of the street organ, progress, as in the 'Complainte des pianos', from romantic dreams to desolation, and emphasize here also that 'la vie est vraie et criminelle'.[36] The poem thus conveys a sense of passing time by means of narrative, and of simultaneity by means of its parallel discourses.

In the 'Complainte des grands pins dans une villa abandonnée' the personified street organ is replaced by the pines, and the personal lament of a disinherited and lonely young man provides the human narrative. The personification of objects that act as witnesses to the human situation is one way of diversifying the *moi*, since Laforgue's voice, already lent to the introductory narrator and to the young man (or girl) who is the main character, has 'become' also the voice of the pines

36 'Complainte des pianos'.

(or street organ). The poem is an amalgam of voices, functioning in relation to each other. It is an amalgam or 'fugue'[37] of moods also, since wit constantly breaks through, tempering the evident pathos. The sun is sulking, the clouds are 'paquets de bitume', the young man will go to Montmartre 'en cinquième classe'. And yet the lament, carried by the moaning of the pines in the wind, has the ability to inspire what Mallarmé, speaking of the street organ, called 'desperate reverie': 'l'orgue de Barbarie, dans le crépuscule du souvenir, m'a fait désespérément rêver. . . . Je la [its crude music] savourai lentement et je ne lançai pas un sou par la fenêtre de peur de me déranger et de m'apercevoir que l'instrument ne chantait pas seul.'[38]

A *complainte* of which we happen, quite exceptionally, to have an early draft, 'Complainte du fœtus de poète',[39] introduces the contrasting strain only in the later version, illustrating Laforgue's inclination to complicate and to 'blur' his original texts. 'Je les retoucherai, je les *noierai*[40] un peu plus', he wrote to his sister: 'La poésie doit être à la vie ce qu'un concert de parfums est à un parterre de fleurs.'[41] As he did this, he was drawing poetry closer to the related art-forms of music and painting, establishing the 'pont mystérieux' of which Delacroix wrote: 'L'écrivain écrit presque tout pour être compris. Dans la peinture il s'établit comme un pont mystérieux entre l'âme des personnages et celle du spectateur.'[42]

The techniques of theatre were useful to Laforgue in this attempt. Two *complaintes* ('Complainte des voix sous le figuier bouddhique', and 'Complainte des formalités nuptiales') employ named characters and, consequently, display considerable formal variety. The former, expanding mingled Buddhist and Hartmannian sentiments, was one that Laforgue considered important, writing to his publisher: 'Une erreur dans cette pièce me désolerait.'[43] Both point forward in their form to *Le Concile féerique*, a drama composed of unpublished poems taken from *Des Fleurs de bonne volonté*.

Techniques drawn from the monologue form developed in the theatre of Coquelin the Younger abound: colloquialisms, the direct address to a supposed spectator, the apostrophe, the aside. One remembers that

[37] Laforgue used the word to describe his poetry: 'les *fugues* vraies | Que crie, au fond, ma riche voix' ('Complainte des complaintes'). He also spoke of 'cette distribution en staccato et en menus enroulements et déroulements *fugués* qui est devenue pour moi un besoin, une envie de femme enceinte dans les vers de rêve' (*LA* 65; my emphases).

[38] 'Plainte d'automne'. [39] *LA* 54–5. [40] Laforgue's emphasis. [41] *OC*, v. 21.

[42] Eugène Delacroix, *Journal*, ed. A. Joubin (Paris, 1980), 29.

[43] *OC*, v. 126.

Mallarmé had originally hoped that 'L'Après-midi d'un faune' might be recited at the Théâtre-Français, and that its second title, after he rejected 'Improvisation d'un faune', was 'Monologue d'un faune'.[44] A number of the personae that Laforgue employs for the self are, as we have seen, connected with legend or folk-song: the Pierrot, the King of Thule, the knight errant, the poor young man, the betrayed husband, the 'blackboulé', or more fantastic versions of the self: the son of Faust, the foetus of the poet, the incurable angel, the poor human body. They are 'characters' who, however strange, are lent familiarity by the definite article that always introduces them. But this familiarity is also belied by the twist that Laforgue gives to the expected personage. We have a knight errant whose only concern appears to be his inability to inspire love, an angel who regrets nothing so much as his purity, a poet-foetus in place of Baudelaire's infant-poet. Antecedents are not necessarily literary. Bizet's *Carmen*, an opera much in vogue throughout Western Europe in 1885, provides the twist in the 'Complainte des blackboulés'. By quoting the opening words of the famous Escamillo/Carmen duet, 'Si tu m'aimes', Laforgue places behind the 'blackboulé', who is nursing a sadistic revenge, the dramatic figure of the betrayed Don José, and thus behind the harsh introspection of an apparently morbid psyche the normalization provided by the legitimate passions of a tragic narrative.[45]

Sometimes the stylized persona is abandoned, and we are given monologues more directly attributable to the poet. The result is not simplification—rather the reverse. The more successful and substantial of these poems illustrate the divisions in the self that will be further developed in the *Derniers Vers*. 'Complainte d'une convalescence en mai', for example, is an occasional poem with a mock-philosophical theme. The tedium of convalescence leads to meditation on life's fundamental problems and to a recognition that the narrator's 'grandes angoisses métaphysiques | Sont passées à l'état de chagrins domestiques'. The fragility of the narrator's affirmations is stressed in a dialogue with the self that employs mobile pronouns, a technique that

[44] 'Les vers de mon *Faune* ont plu infiniment, mais de Banville et Coquelin n'y ont pas rencontré l'anecdote nécessaire que demande le public et m'ont affirmé que cela n'intéresserait que les poètes. J'abandonne mon sujet pendant quelques mois dans un tiroir pour le refaire plus librement plus tard.' (Stéphane Mallarmé, *Œuvres complètes* (Paris, 1945), 1450.)

[45] Laforgue had seen the opera many times. *Carmen* was the Empress Augusta's favourite opera, and in his diary of 1883 Laforgue writes 'Indigestion de *Carmen*' (*AH* 870). Jeanne Bem's subtle 'microlecture' of this poem (in *Hidd. LA* 106–16) does not take into account this obvious (once spotted) allusion, which influences the reader's response to the poem and was presumably intended to do so.

Laforgue uses elsewhere. A 'je' takes issue with a self-addressed 'tu' in free Apollinairean fashion ('Et toi, cerveau confit dans l'alcool de l'Orgueil'), and the two add up to an internal 'nous' ('Nous savons ce qu'il nous reste à faire'). This 'nous' echoes the external 'nous' of the poem's epigraph, the well-known statement referring to Pascal's barbed iron belt: 'Nous n'avons su toutes ces choses qu'après sa mort', an epigraph that, while it indicates that the subject of the poem will be concealed pain, emphasizes also the distance between the Pascalian world and the mundane region inhabited by this poem. The *complainte* employs the alexandrine couplet to combine the presentation of suffering with its subversion ('Convalescence bien folle, comme on peut voir'), while offering on the way a number of bold elliptical formulations: 'Si la Mort, de son van, avait *chosé* mon être', for example, or 'Qui *m'a* jamais rêvé?'[46]

'Complainte d'un certain dimanche', another interior monologue, is an occasional poem whose occasion, the departure of the loved woman, becomes clear only in the third stanza. Before this we have aphorisms about the relations between the sexes, strongly coloured by the end-of-the-affair moment of which we are as yet unaware. Employing the discontinuous technique that Laforgue was to develop further in free verse, it again evokes the disunity of the self. The stanza that introduces the love theme displays three stances towards love in only four lines. We move from apparently bewildered self-questioning to an expression of overstated devotion, and from this to a stock romantic description, the cliché element in which leaves the emotion (such as it is) intact:

> Elle est partie hier. Suis-je pas triste d'elle?
> Mais c'est vrai! Voilà donc le fond de mon chagrin!
> Oh! ma vie est aux plis de ta jupe fidèle!
>   Son mouchoir me flottait sur le Rhin. . .

Shifting stances, again emphasized by a mobile narrative method (the girl is both 'elle' and 'tu'), reflect the frightening inner lack of centre, which is set against the equally alarming anonymity of the external world. Loss of a sense of identity does not imply any absence of feeling, however. Philosophical generalizations and pseudo-conclusions are interrupted by shock expressions of fear or horror:

46 My emphases. Collie reads the second as a misprint for 'n'a', which is to remove its originality.

Que d'yeux, en éventail, en ogive, ou d'inceste,
Depuis que l'Être espère, ont réclamé leurs droits!
O ciels, les yeux pourrissent-ils comme le reste?

When the narrator finally confesses his fear of solitude, which was the theme of many *Sanglot* poems, we find that we have now been offered a landscape in which to situate it, and that it is a nuanced landscape, created from a series of fleeting perceptions—snapshots, as it were, necessarily surrounded by unbridgeable gaps between the pictures, which nevertheless inhabit a recognizable and all too human psychological terrain.

Frequently, the narrator surfaces from these explorations of inner turmoil in a final neat reversal or disclaimer, a technique that we find in the *Derniers Vers*. In this poem it is the shift in the last line from 'Faudra-t-il vivre monotone?' to 'Tâchons de vivre monotone'; in the 'Complainte d'une convalescence en mai' we have the (already quoted) dismissive 'Convalescence bien folle, comme on peut voir'; it is the 'C'était donc sérieux?' of 'Autre Complainte de Lord Pierrot'; the unexpected throw-away, 'Ces êtres-là sont adorables', of 'Complainte sur certains ennuis'; or the Eliotesque ending of 'Complainte des débats mélancoliques et littéraires':

O Hélène, j'erre en ma chambre;
Et tandis que tu prends le thé,
Là-bas, dans l'or d'un fier septembre,
Je frissonne de tous mes membres,
En m'inquiétant de ta santé.

Tandis que, d'un autre côté. . .

The reversals and disclaimers, like the use of the refrain, illustrate the centrality to the volume of techniques of counterpoint. They can be found almost buried in stylistic devices. 'Grande Complainte de la ville de Paris'—a poem that hurls discord and fragmentation at the reader, and is, correspondingly, the only poem in the volume to be written in prose—contains units recuperable as octosyllables and alexandrines, as well as many rhyming effects that set up disturbing echoes of the traditionally rhythmical at the heart of the poem's prose modernity.[47] 'Complainte d'un autre dimanche' uses internal crossed rhyme, as Grojnowksi has convincingly shown,[48] to create the *en abyme* effect of a

[47] P. Collier discusses this poem in *Unreal City*, ed. D. Kelley (Manchester, 1985).

[48] D. Grojnowski, *Jules Laforgue et l'originalité* (Neuchâtel, 1988), 107. Examples of internal crossed rhymes are: 'vent/quand/blancs'; 'aujourd'hui/jalousie'; 'coin/plein'; 'arbres/marbres'; 'fait/après'.

poem within a poem. Contrapuntal links are formed between poems; this poem, for example, stands in apposition to its predecessor, 'Complainte d'un certain dimanche', confronting and reflecting it. The same can be said of the two Lord Pierrot poems. But chiefly, of course, systems of opposition and relationship function between the parts of the poem. The narrator's personal lament in 'Complainte d'un autre dimanche' is set against the apparent objectivity of a precisely observed pictorial setting that turns out to be an accurate reflection of the narrator's internal landscape. The transience of the former, emphasized by first and last lines (*C'était* un très-au vent d'octobre paysage'; 'Ce *fut* un bien au vent d'octobre paysage. . .'[49]), echoes the human instability that, as we have seen, is the true subject of the poem.

Ultimately, despite the scope it offered for the portrayal of a range of different and changing voices, Laforgue began to find the *complainte* form limiting rather than liberating.[50] It had enabled him to discover and explore a number of ironic distancing techniques, but the many voices had become, self-confessedly, 'gerbes . . . d'un défunt moi'.[51] He moved on, at first to the still greater artifice of *L'Imitation de Notre-Dame la lune*, and finally, in the *Derniers Vers*, to an abandonment of formal stylization—and even of versification—in his constant attempt to record the impact of reality, or, in Gide's later words, to present 'la rivalité du monde réel et de la représentation que nous nous en faisons'.[52]

The aim pursued in the *Complaintes* necessarily involved breaking some of the venerated rules of French versification and, therefore, adopting a *vers libéré*. This might be thought natural in any case in a volume that derived its inspiration from folk-song. But the same critics who found the work intolerably obscure no doubt considered that Laforgue, like Donne before him, 'for not keeping of accent deserved hanging'.[53] Laforgue's handling of rhyme came in for particularly vehement treatment.[54] One critic had the wit at least to realize that the transgressions were intended: 'Il est évident qu'on dira des *Complaintes* que ce n'est pas rimé suivant les règles données par le maître Banville.[55]

[49] My emphases.

[50] *LA* 83: 'car je voudrais faire autre chose que des *Complaintes*.'

[51] 'Complainte des complaintes'.

[52] A. Gide, *Les Faux-monnayeurs*, sect. II, ch. 5.

[53] Said by Ben Jonson, and reported in *Notes on Ben Jonson's Conversations with William Drummond*, ed. C. H. Herford and Percy Simpson (London, 1937), i. 133.

[54] The reviews are printed in Debauve, *Laforgue*, 193–215.

[55] T. de Banville, *Petit Traité de poésie française* (Paris, 1872).

Mais cette indépendance prosodique a au moins cela pour elle, qu'elle n'est pas le résultat de l'impuissance. C'est voulu.'[56]

Laforgue insisted that his poems were 'rimées à la diable',[57] and by this he meant that he had sought out bold and interesting rhymes and had neglected convention. Instead of alternating masculine and feminine rhymes, 'Complainte d'un autre dimanche' uses feminine rhymes throughout, a technique that Verlaine had pioneered for effects of delicate evocation. Laforgue rhymes for the ear and not the eye. He seems actually to cultivate the juxtaposition of singular and plural rhyming words that he had recommended to Mme Mültzer already in 1882.[58] *Rimes pauvres* are common, but so are intensive rhyming effects, created by repeating a rhyme three or more times, often in complex patterns related to the formal structure of the poem. ('Complainte de la bonne défunte' uses the same two feminine rhymes throughout the entire poem; 'Complainte des printemps' uses the same feminine rhyme in the poem's quatrains only; 'Autre Complainte de l'orgue de barbarie' uses each rhyme in its main stanzas three times, the refrain rhyme six times.) Experimentation is constantly in evidence. In 'Complainte des pubertés difficiles' one rhyme is regularly carried over from one stanza to the next, reducing the separateness of stanzas. In 'Complainte de la fin des journées' the refrain uses the same rhymes throughout, and is thus set apart from the main stream of the poem.[59] In 'Complainte de l'ange incurable' the rhymes of the first two couplets are regularly repeated in the poem's short refrain, and act as wan echoes of the main text.[60] Dissonances occur in place of, and in addition to, rhymes, as in the complex patterning of 'Complainte du roi de Thulé',[61] and can result, as here, in a sense of subtle musicality, caused by the combination of rhyme and the related dissonance, of repetition and a variation on it. Rhyming on proper names or on words drawn from foreign vocabularies (and sometimes on both at once) lends the rhyme exoticism or humour; as in 'Missouri' and 'Paris' ('Complainte de la lune en province'), 'draps' and 'Léda' ('Complainte de Lord Pierrot'), or 'affiche' and 'sandwiche' ('Complainte du pauvre chevalier-errant'). The 'sans-gêne' of Laforgue's rhymes, the quality he twice

[56] Debauve, *Laforgue*, 199. [57] *AH* 810. [58] Ibid. 792.

[59] This pattern, with variations, is adopted in a number of poems, including the highly structured 'Complainte des pianos' and 'Complainte des grands pins'.

[60] *a*: 'cendres/tendres'; *b*: 'd'antan/étang'; *refrain*: 'Et vous, tendres | D'antan'.

[61] *1st stanza*: 'Thulé/immaculé/palais'; *2nd stanza*: 'allait/clés/lait'; *3rd stanza*: 'ourlé/Thulé/ululait'; *Final stanza*: 'lait/clés/Thulé/immaculé'.

praised in Kahn's,[62] is one of their most engaging characteristics, as can be seen in the following:

Les fiords bleus de la Norwège
Les pôles, les mers, que sais-je?
('Complainte de la lune en province')

Si tu savais, maman Nature,
Comme Je m'aime en tes ennuis,
Tu m'enverrais une enfant pure,
Chaste aux *"et puis?"*
('Complainte-Placet de Faust fils')

or in the more provocative lines of 'Complainte d'une convalescence en mai', with their humorous hesitancy:

Je ne veux accuser personne, bien qu'on eût
Pu, ce me semble, mon bon cœur étant connu. . .

Laforgue's sensitivity to the weighting of syllables and to relative degrees of stress leads to a range of novel rhythmical effects, which we can be sure were calculated. Short lines, like stanzas, are contrasted with long, the caesura is eradicated or the line disrupted by unexpected or frequent breaks, the *impair* is used for effects of disharmony and ambivalence, notably in the hendecasyllable of 'Complainte du fœtus de poète', with its constantly shifting main stress. In 'Complainte du pauvre chevalier-errant' Laforgue created what he called a 'strophe absolument inédite à vers de 14 pieds',[63] which he asked Henry to emphasize in the review that he wrote of the *Complaintes*. These experimental rhyming and syllabic combinations formed part of Laforgue's attempt to make details of technique relate to meaning. 'Que pensez-vous du vers de onze pieds?', he wrote to Kahn, 'et par la même occasion, que pensez-vous aussi de l'infini?'[64] As a result, even his experimental lines have an integrity such as might be expected from more predetermined forms. A more rigid approach towards prosody would have seemed to him, paradoxical as this may appear, a betrayal of the artistic purity to which he aspired.[65] He set out the position most

[62] *AH* 803, 852.

[63] *OC*, v. 132. In the same letter to Henry he spoke of 'mes rythmes et rimes absolument inédits, ce qu'il n'y a pas eu depuis quinze ans, nulle part, sauf en Verlaine un peu' (ibid. 132).

[64] *LA* 55.

[65] 'Je sais qu'en quatre ans je pourrai faire fortune si je voulais écrire des romans à la Guy de Maupassant. *Bel Ami* est d'un maître, mais ce n'est pas de l'art pur. Peut-être ce désir de créer de l'art pur est-il un louable mais pauvre désir de nos vingt-cinq ans?' (*OC*, v. 142.)

directly when he defended himself against accusations of imitating Corbière: 'Corbière ne s'occupe ni de la strophe ni des rimes (sauf comme un tremplin à concetti) et jamais de rythmes, et je m'en suis préoccupé au point d'en apporter de nouvelles et de nouveaux. J'ai voulu faire de la symphonie et de la mélodie.'[66] The 'melody' involved using a vocabulary that frequently appeared unpoetic—familiar and tongue-in-cheek—when this contributed to the particular poetic effect that he desired. It involved shocking his critics by rhyming 'Saint-Malo' with 'sanglots',[67] and 'bocks' with 'coq'.[68] It involved using varied and eccentric syllabic lengths and intricate contrapuntal patterning, so that the poetic line, far from being a mechanical unit, became the base for a series of live combinations. Being concerned with 'stanzas and rhymes' meant the creation of new effects, rather than adherence to the formulations of the past, offering the reader a constant interplay between the traditional and the original, the expected and the novel. Laforgue was no doubt in agreement with (and perhaps behind) these sentences from Henry's review of the *Complaintes*: 'M. Jules Laforgue fut ainsi conduit à un genre de composition où la tenue prosodique conventionnelle n'est pas de rigueur. De là ces complaintes . . . entrelacis de notes perpétuelles, échos d'humour de belle race, trouvailles de formules, bouquet de rythmes et de rimes dont la variété réjouit le savant parfois inquiet du nombre.'[69]

[66] Ibid. 137.
[67] From 'Complainte de la vigie aux minuits polaires'.
[68] From 'Complainte de l'oubli des morts'. The anonymous critic of the *Paris illustré* of 1 Oct. 1885 wrote: 'C'est un recueil de poésies bizarres, où Saint-Malo rime avec sanglots et bocks avec coq.' (Debauve, *Laforgue*, 207.)
[69] Debauve, *Laforgue*, 195.

3

L'Imitation de Notre-Dame la lune: 'Le sang-froid du dilettantisme qui passe'

PIERROT AND THE MOON

LAFORGUE turned briefly from the exploration of multiple personae to follow a popular cult—that of moon-worship. One might say that it was two popular cults, since his volume has two subjects: the moon and her worshipper, the Pierrot. Laforgue knew perfectly well that neither subject was original. He was offering nothing more than 'quelque chose comme *contribution* . . . au culte de la lune', containing 'plusieurs piécettes à la lune'.[1] He wrote the poems fast, which was not unusual, since he worked in bursts of enthusiasm and there were long periods when he did other things—art criticism or novel-writing. (The novel *Saison*,[2] which he planned at this time, was described by Téodor de Wyzewa as 'du Stendhal gracieux'.) But what was unusual was the flippancy with which he referred to the volume. It was clearly intended from the outset to be a minor volume—a brief respite from his main preoccupations; the poems 'piécettes', the whole a mere 'contribution'. With this dismissive tone went a note of artistic confidence. It is in fact a volume in which technical control shines through many of its witty and elegant poems. At the time that Laforgue composed it he was also writing his early *Moralités légendaires*, and he commented significantly: 'Qu'il est plus facile de tailler des strophes que d'établir de la prose!'[3] The poems of the volume are very distinctly 'taillés', and the assured tone represents an attitude that was soon to be superseded. It was, as Laforgue described it in a prose fragment, 'le sang-froid du dilettantisme qui passe'.[4]

1 *LA* 100. Laforgue's emphasis.

2 The title is mentioned in Apr. 1885 (*OC*, v. 120). April, May, and June seem to have been largely devoted to the work. Jean-Aubry gives us the following information: 'Il dut détruire ce qu'il en avait fait: il ne nous en est guère parvenu plus que le titre. Téodor de Wyzewa qui en avait vu davantage déclarait que c'était "du Stendhal gracieux"'. (*LA* 110.)

3 *OC*, v. 128.

4 *MP* 56–7. The MS of the fragment quoted here without a date, which was once part of the Jean-Aubry collection, begins 'dimanche-soir 10h. 19 juillet 85'.

But the volume was original in a variety of ways. Laforgue's Pierrot is a highly personal adaptation; his moon, as a contemporary reviewer pointed out, was like none of its predecessors,[5] and the volume's particular brand of moon-worship combined the Pierrot figure and the moon in an unprecedented way. For the *Lunes* are structured around this relationship. Moon poems flank two sets of Pierrot poems; the Pierrot figure is obliquely introduced in the last of the introductory moon poems, and the moon is reintroduced in the last of the Pierrot poems.

Part of the subtlety of the volume depends on its element of religious parody. The moon, a female figure, is inspired in part by *Salammbô*'s Tanit.[6] Ambivalence is suggested at the start by the grotesque opening denunciation of the sun, symbol of life, a denunciation paralleled by the overstated description of the barren craters, rigidified flora and fauna that make up the moonscape of death. The human figures, narrator and Pierrot, plead, worship, prostrate themselves before a largely silent divinity, and aspire to a loss of consciousness in a kind of Hartmannian nirvana. Significantly, as in Mallarmé's 'Hérodiade', negation evokes also its opposite principle, and the pursuit of a dandified ideal of renunciation conjures up the landscape of frustrated desire.

The moon is given two faces, human and natural; that of the sophisticated but sterile coquette—far removed from Baudelaire's sensuous *bienfaitrice*[7]—and a barren landscape, described in modern scientific and decadent vocabulary (Laforgue had read Huysmans's *A rebours*, published the year before[8]). One of the few reviews that the volume received concentrated largely on the decadence.[9] The bombastic shock images, juxtaposing concrete and abstract, modern and ancient, natural and man-made, still surprise: 'palmiers de corail blanc aux résines d'acier', 'geysers de mercure', 'champignons aménagés comme des palais', 'lys marmoréens à sourires hystériques', 'nébuleuses d'occasion'.[10] This is not mere verbal bravado, however, for, like the Pierrot poems, it rests on a psychological foundation, resembling Laforgue's more successful exploration of the aquarium as a metaphor for the subconscious.[11] On a grander scale, but in less self-conscious

[5] The review is published in full in J.-L. Debauve, *Laforgue*, 218.

[6] The volume is dedicated to 'la petite Salammbô, prêtresse de Tanit', as well as to Kahn.

[7] Charles Baudelaire, 'Bienfaits de la lune', *Le Spleen de Paris*.

[8] *OC*, v. 89.

[9] Debauve, *Laforgue*, 218–19.

[10] 'Climat, faune et flore de la lune'.

[11] Prose descriptions of the aquarium were incorporated by Laforgue into 'Salomé'. Grojnowski provides variants of these passages in *ML* 200–7.

manner, Zola's symbolic treatment of the geological in the infernal underground regions of *Germinal*, a work published in the same year, can be seen to be undertaking something parallel. Psychological fears and inhibitions, translated into images of negativity, stasis, and renunciation, are the subject of both moon and Pierrot poems: the two are complementary, and what is humorously described as 'ce Port-Royal d'esthètes'[12] anticipates and reflects its officiating priest.

'Adorez-vous le cirque?', Laforgue wrote to Mme Mültzer in 1882. 'Je viens d'y passer cinq soirées consécutives. Les clowns me paraissent arrivés à la vraie sagesse. Je devrais être clown, j'ai manqué ma destinée.'[13] And, of course, the artist as clown, while it is an image that pervades all literature, haunted the French nineteenth-century imagination in particular.[14] The famous mime Deburau fixed its image as that of the Pierrot, and numerous devotees celebrated it in verse; Gautier, Banville, and Verlaine among them. It is perhaps only in the late twentieth century that it has lost some of its symbolic appeal. Estragon and Vladimir, originally conceived by their creator as clowns, have, with Beckett's reluctant permission, become tramps.

Robert Storey examines the figure psychoanalytically,[15] seeing in it an image of infantile regression. There can be no doubt that, as such, this particular 'autoportrait travesti'[16] is never likely to depart from the literary scene. Its appropriateness to Laforgue, a classic sufferer from the void created by the absence of the mother-figure, is clear. In some sense, the Pierrot must always be a 'void' to be filled by each artist in a different manner, and no doubt it owed some of its success at this period to its malleability to a range of private imaginings. Laforgue's Pierrots, for they are a band, are initially described in terms that make them appear bizarre. But beneath their artificial pose lies a sensitivity that normalizes them, exposing cynicism for the pretence that it is. The

[12] 'Un Mot au soleil pour commencer'.

[13] *AH* 753.

[14] The area is explored in J. Starobinski, *Portrait de l'artiste en saltimbanque* (Geneva, 1970), and in R. Storey, *Pierrots on the Stage of Desire* (Princeton, NJ, 1985). The figure of the Pierrot, derived from the Pedrolino of the *commedia dell'arte*, took on in the 18th century some characteristics of the French clown Gilles, who was immortalized by Watteau, and represented in the 19th, in Gautier's words, 'l'âme humaine encore innocente et blanche, tourmentée d'aspirations infinies vers les régions supérieures' (Starobinski, *Portrait de l'artiste*, 8). Even Flaubert tried his hand at the theme in *Pierrot au sérail*. Starobinski sums up the situation thus: 'Depuis le romantisme . . . le bouffon, le saltimbanque et le clown ont été les images hyperboliques et volontairement *déformantes* que les artistes se sont plu à donner d'eux-mêmes et de la condition de l'art.' (Starobinski, *Portrait de l'artiste*, 9.)

[15] Storey, *Pierrots*.

[16] Starobinski, *Portrait de l'artiste*, 9.

mask is therefore valuable to a poet whose irony is intended to allow sentiment indirect expression. Through it, Laforgue can suggest the idealism of his 'suraigu' period without being committed to supporting it. The balanced oppositions on which the poems centre: the ideal and the real, cynicism and vulnerability, the intellect and the emotions, are central to any description of human nature, and this is disingenuously pointed out when the Pierrot, identifying with humanity, says:

> Ces gratuites antimonies
> Au fond ne nous regardent mie.[17]

The bold assertion does not convince. 'Le beau rôle' may be 'de hausser à tout les épaules', but this is a mask, achieved with effort ('dût on se battre les flancs').[18] Duality is inescapable. And so, we have the famous 'rire aux déchirantes gammes' of 'Pierrots IV'; that 'alliance', in T. S. Eliot's words, 'of levity and seriousness (by which the seriousness is intensified)'.[19] 'In Dryden', Eliot wrote, 'wit becomes almost fun, and thereby loses some contact with reality; becomes pure fun, which French wit almost never is.'[20] Laforgue's Pierrots, who contrast with Deburau's simpler impersonations of the unsuccessful romantic lover, do not lose sight of reality, because, despite their distanced pose, reality provides both starting- and finishing-points. The world of chaos and disorder may appear to be left far behind, yet, in the deftest manner, the poems lead us back to it. The Pierrot's victory is a victory of style. The poems construct an artificial world: all realistic precision of circumstance is removed. The Pierrot as philosopher, as artist, not only conquers his vulnerability, but rejects life for a more elevated goal. His private and solipsistic world is transcended and universalized. But the pose, which in simple psychological terms is a mechanism for survival, provides in inverse manner what Joyce termed the 'technical bridge' that allows the unconscious expression, so that the reader is simultaneously led back towards the uncharted and disturbing psychological area from which the Pierrot claims to escape. If the first satisfaction that the poems offer is intellectual—the supposed victory of art over life, of self-sufficiency over desire—the final satisfaction is emotional—and serious. The balancing act is beautifully maintained. Artifice wins the day, and yet these are not poems about artifice.

[17] 'Pierrots V'.
[18] Ibid.
[19] T. S. Eliot, *Selected Essays* (London, 1932), 296.
[20] Ibid. 301.

'L'AUTOPORTRAIT TRAVESTI'

The dramatic highpoint of the volume consists of two supposed dialogues between Pierrot and 'Elle', illustrations of love's impossibility, each of which is flanked by a series of Pierrot poems, the first a depiction of the Pierrot figure, the second a series of short soliloquies. The first of these poems is subtitled 'On a des principes'; the second 'Scène courte mais typique'. Both are dramatic monologues. Laforgue may have been influenced by Charles Cros's humorous prose monologues, performed by Coquelin the younger at the Comédie-Française; or by Huysmans's and Léon Hennique's *Pierrot sceptique*, which delighted Parisian theatre audiences in 1881 and which Laforgue recommended to Charles Henry in book form in the following year.[21] After the lifting of the prohibition on speech at the Théâtre des Funambules in 1865, writers had taken up the challenge of finding words for what had previously been expressed in mime. Pierrot was, of all characters of mime, by definition the most complicated and inward, irony being written into a renunciation that masked a secret desire for fulfilment.

The first of these two central poems is an interior monologue of an unexpected kind. The woman's protestation of love ('Je t'aime pour toi seul') is disbelieved and mocked by the Pierrot. The mockery continues even after the girl has died—presumably of grief—and can be explained only as the revenge of the imagination on a reality of a very different kind—namely, the girl's actual indifference. The poem can thus be read as a faithful transcription of a wish-fulfilment fantasy: the underlying vulnerability of the poet is not revealed until the last line, when the tone is reversed by the admission of feeling. The dramatic monologue has a complex structure: it leads off with the girl's quoted and ridiculed speech. The tone is light, the language colloquial, the mocking refrain 'Oh! là, là' occurring in each of the first three stanzas. The girl, who was described as a 'character' in the third person in the first half of the poem, becomes, as the tension increases, the addressee of the second half. We thus have a shift of narrative stance, which both forces the reader to view the situation objectively and invites his emotional involvement.

The dramatic element is central also to the next Pierrot poem, subtitled 'Scène courte mais typique':

[21] *AH* 758.

Il me faut vos yeux! Dès que je perds leur étoile,
Le mal des calmes plats s'engouffre dans ma voile,
Le frisson du *Vœ soli!* gargouille en mes moelles. . .

Vous auriez dû me voir après cette querelle!
J'errais dans l'agitation la plus cruelle,
Criant aux murs: Mon Dieu! mon Dieu! Que dira-t-elle?

Mais aussi, vrai, vous me blessâtes aux antennes
De l'âme, avec les mensonges de votre traîne,
Et votre tas de complications mondaines.

Je voyais que vos yeux me lançaient sur des pistes,
Je songeais: Oui, divins, ces yeux! mais rien n'existe
Derrière! Son âme est affaire d'oculiste.

Moi, je suis laminé d'esthétiques loyales!
Je hais les trémolos, les phrases nationales;
Bref, le violet gros deuil est ma couleur locale.

Je ne suis point 'ce gaillard-là!' ni Le Superbe!
Mais mon âme, qu'un cri un peu cru exacerbe,
Est au fond distinguée et franche comme une herbe.

J'ai des nerfs encor sensibles au son des cloches,
Et je vais en plein air sans peur et sans reproche,
Sans jamais me sourire en un miroir de poche.

C'est vrai, j'ai bien roulé! j'ai râlé dans des gîtes
Peu vous; mais, n'en ai-je pas plus de mérite
A en avoir sauvé la foi en vos yeux? dites. . .

—Allons, faisons la paix, Venez, que je vous berce,
Enfant. Eh bien?
—C'est que, votre pardon me verse
Un mélange (confus) d'impressions. . . diverses. . .

(*Exit*)

Here the two 'characters' are, as we would expect, sharply contrasted. As in the earlier poem, the woman is seen as frivolous, worldly, deceitful, given to false sentiment: the Pierrot is honest, sensitive, instinctively idealistic, but forced into cynicism by experience. He faces a genuine impasse. He can neither accept the woman's offered love, nor be happy without it. His behaviour is bewildering to the woman, whose statements are interpreted by the Pierrot as insincere, though they are not necessarily so. Male and female attitudes are highly stylized and are presented for the greatest possible divergence, as is natural in a poem whose theme is the impossibility of mutual understanding between the sexes. But this stylization is also a faithful representation of the workings of the inner mind, setting out the conflict for, and to, its own satisfaction. The Pierrot's speech is, of course, not an actual speech, but the mental rehearsal for a speech—in fact, despite the subtitle 'Scène', just as much an interior monologue as was the preceding poem. It is an accusation and a self-defence—a marshalling of the evidence for and against, with the dice heavily loaded, naturally, against the girl.

In contrast to the earlier poem, the girl is the direct addressee throughout the poem, and appears as a 'character' only in moments of reflection, which have the function of theatrical asides. The asides, which create a context for the action of the poem, are woven into the dramatic narrative, and, since they often offer conflicting 'aperçus', the poem invites a range of responses from the reader, from emotional participation when the girl is addressed ('il me faut vos yeux!'), to the analytical detachment necessary to appreciate the destructive wit of such a statement about the girl as: 'Son âme est affaire d'oculiste'. By an imaginative reconstruction, the reader must follow the narrator in his dual role of participant and spectator. The rift between intellect and feeling is made abundantly clear, but, neither analysed nor commented on, it is offered to the reader in its raw state.

This is achieved only with much authorial artifice. Here the Pierrot's discomfiture is mocked by the fact that 'agitation' must be counted as five syllables instead of the expected four. In the following stanza five syllables again are needed in the word 'complications' for the anticipated alexandrine, and thus the Pierrot's irritation with the worldly trivia that make up the girl's life is emphasized and ridiculed. A third, similarly humorous, effect occurs at the end of the poem:

> —C'est que: votre pardon me verse
> Un mélange (confus) d'impressions. . . diverses. . .

Here the lengthening of the word 'impressions' is further stressed by suspension points. The Pierrot's final mood is very far from the unadulterated happiness that might have been foreseen as the natural result of the girl's pardon. The four-syllabled 'impressions' forces the reader to linger on this irony.

Similar examples of wit depending on the reader's metrical expectations can be seen in the first of the poems discussed: the humorous hesitancy, matching the preposterously flippant tone of the Pierrot, that is suggested by ending the lines on the words, 'troisième' and 'moins' rather than on 'jour' and 'odeur':

On sait que tu dois ressusciter le troisième
Jour, sinon en personne, du moins

Dans l'odeur, les verdures, les eaux des beaux mois!

Such essentially whimsical line endings, perhaps inspired by Verlaine's *Fêtes galantes*, are not unusual:

Je me console avec la
Bonne fortune
('Locutions XI')

Ne leur jetez pas la pierre, ô
Vous qu'affecte une jarretière!
('Pierrots IV')

and a playful wit, seen also in intensive rhyming effects such as the following, is central to this volume:

Cautérise et coagule
En virgules
Ses lagunes des cerises
Des félines Ophélies
Orphelines en folie.
('Stérilités')

Farce was inseparable from the theatrical figure that Laforgue chose for his disguised self-portrait. His Pierrot is a jester and a poseur. He expostulates, exclaims, commands, questions, pleads, changing his narrative stance with bewildering speed. The reader is addressed and called in to judge, and the girl who has been the addressee ('vous' or 'tu') becomes a character ('Elle' or 'la femme'). At times, reader and speaker are united in the inclusive 'nous' of the Pierrots. But the two have been brought together in more than form of address, for the reader is forced to admit that these initially odd figures have become disarmingly acceptable—indeed, recognizable for their measure of common humanity. They have come to represent the sensitive person's response to life's vulgarities and inadequacies. Beneath a humorous mask ('un air d'hydrocéphale asperge'),[22] Laforgue has lent the Pierrot figure his sensitivity and his rigorous intellectualism. The whole is a virtuoso exercise, and artifice and provocation are natural and basic ingredients. Laforgue promises that he will be more honest with his reader in future, that he will write 'un livre enfin de bonne foi',[23] but he does this in a

22 'Pierrots I'. 23 'Avis, je vous prie'.

final poem of doubtful seriousness that includes a contrived eleven-syllable alexandrine designed to shock his critics.[24] A similar message was conveyed to Bourget with more respectful seriousness: 'Ceci n'est qu'un *Intermezzo*. Attendez donc et donnez-moi jusqu'à mon prochain livre.'[25] To the late twentieth century, the exploration of the 'néant' within and without ('ne rien dire pour dire le rien',[26] as Grojnowski styles it) may seem more serious than it no doubt did to Bourget.

Ezra Pound appreciated the 'force' that was allied to 'désinvolture' in Laforgue's works ('why such power should coincide with so great a nonchalance of manner'), and to illustrate the point he chose the tenth of the Pierrots' sixteen *locutions*:

Que loin l'âme type
Qui m'a dit adieu
Parce que mes yeux
Manquaient de principes!

Elle, en ce moment,
Elle, si pain tendre,
Oh! peut-être engendre
Quelque garnement.

Car on l'a unie
Avec un monsieur,
Ce qu'il y a de mieux,
Mais pauvre en génie.

'Laforgue is perhaps incontrovertible', he wrote; 'John B. Yeats has written of the relation of art and "certitude" and we are perhaps too prone to connect "certitude" with the "strong silent man" of the Kinema. There are, however, various species.'[27] All species, in Pound's eyes, introduce change. After Laforgue, certain ways of writing were no longer possible. 'I think', he wrote, 'that he marks the next phase after Gautier in French poetry.'[28] This is clearly true, but it is interesting

[24] L. G. Mostrailles, in *Lutèce* 209, 15-29 Nov. 1885, responded suitably: 'Vanier, dont *Lutèce* publie plus haut une protestation indignée, vient de faire paraître une nouvelle plaquette de M. Laforgue dont je ne veux retenir que ces deux alexandrins, dont un de *onze* pieds, extraits de la pièce finale: "Oh! laissez-moi seulement reprendre haleine, | Et vous aurez un livre enfin de bonne foi . . ." On ne peut pas dire d'une façon plus explicite que les *Complaintes* et l'*Imitation de Notre-Dame la lune* sont deux livres de fumisterie pure; c'est très bien.' Reprinted in Debauve, *Laforgue*, 216.

[25] Written on the copy that Laforgue gave to Bourget.

[26] D. Grojnowski, *Jules Laforgue et l'originalité* (Neuchâtel, 1988), 127.

[27] E. Pound, 'Irony, Laforgue and Some Satire', in *Literary Essays of Ezra Pound* ed. T. S. Eliot (London, 1954), 283-4.

[28] E. Pound, Ibid, 282.

that Pound makes his point with so slight a piece—and any of the Pierrots' *locutions* would have served him equally well. The choice is, in fact, an appropriate one, because the poem is both like the Gautier of *Émaux et camées* in its wit and delicacy, and quite different in its aims and effect. Gautier's plastic poetry is concerned with the creation of a state of heightened beauty above the randomness of life. Beauty is truth: for Laforgue, truth may be beauty. The Pierrot, removed though he may be from the world of human emotions, reflects this world. The stylized *Lunes* linger in the mind for a psychological quality that is quite foreign to Gautier.

4

Des Fleurs de bonne volonté: 'Bonne girouette aux trent' six saisons'

THE HAMLET BACKCLOTH TO THE *FLEURS*

BEFORE treating the 'livre de bonne foi' that followed the *Lunes*, it seems appropriate to consider the figure that stands behind this volume, and which contrasts sharply with the Pierrot. It is, of course, Shakespeare's Hamlet, but it appears only obliquely by means of quotations from the play, used as epigraphs to many of the individual poems and to the volume as a whole. *Hamlet* offers a love scenario that is no less negative than that of the Pierrot, but this is its main point of comparison with the hero of the earlier volume, for to move from the 'double grimaçant' of mime to Shakespeare's 'noble prince' was to move from one literary extreme to its opposite. The mockery is now in reverse. If Laforgue as Pierrot diminishes the author, Laforgue as Hamlet is unacceptably pretentious. The insignificance of the authorial figure is highlighted, and his aspiration to a stature comparable to that of the tragic hero ridiculed.[1]

What is more, the tragic fate itself is questioned. In this volume Laforgue is concerned both with the sense of impossibility and with tentative exploration. Beneath objections and difficulties, a possible escape route is being sought towards some kind of salvation through love. This direction is seen more strongly in the later poems, and it is one of their strikingly original features. Laforgue turns to a future not of vague reverie, such as that found in Verlaine, but one based on realistic awareness. His modernity lies in part in his assault on an intrinsically unpoetic area, a region between Baudelaire's extremes of mud and gold, for which a poetic voice had still to be invented. The poet's task is to find a language *for* experience, Helen Vendler rightly argues.[2] Laforgue

[1] The resemblances between the two figures have been stressed by critics: 'Pierrot's theatrical and literary history is the record of his vacillations between two dramatic and psychological "types". At one point stands his Italian predecessor Pedrolino . . . At the other stands Hamlet.' (R. Storey, *Pierrots on the Stage of Desire* (Princeton, NJ, 1985), 77.) The two figures are linked, although negatively, in Stéphane Mallarmé's sonnet 'Le Pitre châtié' in the phrase 'reniant le mauvais Hamlet'.

[2] H. Vendler, *The Music of What Happens* (Cambridge, Mass. and London, 1988).

at this point, consciously or unconsciously, set himself this task, and the result is a volume in which the tone of the poems and their metres vary widely, an experimental and quickly outgrown volume, which nevertheless paved the way for the highly original *Derniers Vers*.

What the epigraphs from *Hamlet* chiefly provide is a climate in which to read the poems. Sometimes, it is true, they have a particular appropriateness to the poem they introduce. They may have a 'riddle-solving' function, as in 'Il pleut sans but', where the epigraph clinches the idea of an unwanted pregnancy that the poem suggests. Or they may show the poem to be allegorical. Prefixed by Hamlet's exchange with Ophelia (Ophelia: ''Tis brief my lord'; Hamlet: 'As woman's love'), the physical storm that 'Aquarelle en cinq minutes' describes cannot be read as anything but a metaphor for human love. The epigraph, standing in apposition to the text, has become a technique for humorous allusion. But usually its function is more serious and less particularized. In his manuscript versions of the poems Laforgue frequently transposed quotations from one poem to another, juggling, as it were, with their suitability to a particular text. More important, one supposes, than the precise aptness of a quotation to an individual poem was the atmosphere provided by Hamlet's unavailing love for the equally helpless Ophelia. Against this tragic backcloth, the mundane patterns of failure and the tentative patterns of hope of ordinary love could be explored.

Hamlet's popularity in French nineteenth-century artistic milieux is reflected in Delacroix's famous lithographs, which at one time decorated Baudelaire's walls. Mallarmé is perhaps the most famous of Laforgue's contemporaries to be haunted by the figure of Hamlet 'lisant au livre de lui-même'.[3] Yves Bonnefoy traces a thread of Hamletism from Baudelaire to Laforgue, and suggests that the title of Baudelaire's poem 'La Béatrice', in which the poet refers to himself as 'cette ombre d'Hamlet', should have been 'Ophélie': 'A-t-on remarqué, d'ailleurs, que cette "Béatrice" qui apparaît ainsi au seuil du poème aurait dû bien plutôt, devant cette "ombre d'Hamlet", être nommée Ophélie?'[4] Logically, he is of course correct. But 'Béatrice'—the original title of two other poems by Baudelaire—is appropriate to Baudelaire as Ophelia would not have been. Beatrice has unparalleled heroic stature as an eternal female ideal. The frail, immature, Northern, rather than

[3] Stéphane Mallarmé, *Œuvres complètes* (Paris, 1945), 1564. For a description of the French response to the figure of Hamlet, see Helen Phelps Bailey, *Hamlet in France from Voltaire to Laforgue* (Geneva, 1964).

[4] Y. Bonnefoy, 'Hamlet et la couleur', in *Hidd. LA* 169.

Mediterranean, figure of Ophelia, whose evocative force lies in her innocence rather than in her virtue, in her sexual inexperience rather than her spirituality, was discovered for the French by Laforgue. Her death had already been the subject of paintings and poetry. Laforgue, more than anyone else, gave her life mythical status.

The passages selected for quotation are mainly fragments of dialogue between Hamlet and Ophelia, taken from the 'play' scene and their last scene together. They illustrate her incomprehension and sexual vulnerability, but the emphasis lies with his feigned madness, his false gaiety, his mockery and rejection of her. Inevitably, cruelty, pathos, and a sense of betrayal are central. Elements from his 'Get thee to a nunnery' speech introduce three poems. The volume as a whole is prefixed by the opening line of Ophelia's famous lament: 'O, what a noble mind is here o'erthrown!', and after this is set an abbreviated version of Hamlet's dying statement to Horatio:

Had I but time!
O, I could tell you,—
But let it be.

Tragic incomprehension is thus the expected central theme of the volume, and in a range of tones, though never without some irony, this is what we find. Unlike Hamlet, however, the authorial figure *does* have time, and an ambiguous future opens out beyond the shadow of tragedy.

What did the figure of Hamlet mean to Laforgue? To answer this, we must turn to his short story 'Hamlet ou les suites de la piété familiale', which he had finished by August 1885.[5] Attracted by Hamlet's sensitivity, which he shares, by his essentially artistic 'inability to live', Laforgue makes of him a parody of the decadent aesthete, pursued by ennui, in the manner of Huysmans's Des Esseintes, the hero of *A rebours*. Laforgue's narrator is, however, more detached than Huysmans's. Far from being exemplary, the tale is parodic. Laforgue knows his hero to the core, and is exposing him. The reader meets a symbolic landscape on the first page: the 'tour d'ivoire' in which Hamlet hides from the world, the stagnant creek that mirrors his soul. Elements of grandeur in the Shakespearian play are systematically

[5] For an analysis of this work from the angle of parody, see Michèle Hannoosh, *Parody and Decadence: Laforgue's 'Moralités légendaires'* (Columbus, Oh., 1989). The complexity and interest of the *moralité* emerge strongly also in Peter Brooks's analysis of the Hamlet figure in France: 'The Rest is Silence: Hamlet as Decadent', in W. Ramsey (ed.), *Jules Laforgue: Essays on a Poet's Life and Work.*

undercut and devalued. The former king, Hamlet's father, is a vulgar sexual adventurer; the queen is not his mother at all: he is the son of a gipsy girl who died at his birth. This makes him the half-brother of the clown Yorick. One of Hamlet's occupations is to stick pins in the effigies of Gerutha, the queen, and the usurper king, Fengo, names taken from the earliest version of the Hamlet story.[6] At the start of the *moralité* Ophelia has quite simply 'disappeared': perhaps she has gone home to her relatives in the country, in a fit of pique, it is suggested. Hamlet is a prince more interested in the aesthetic merits of the play he has just written, supposedly to 'catch the conscience of the King', than in his proper mission, the avenging of his father's murder. He falls in love with the actress Kate (a doubtless intentional merging of Shakespearian characters, of comedy and tragedy), and elopes with her. Happening to pass near Ophelia's grave, he visits it on an impulse and is murdered there by Laertes, the hero who, unlike Hamlet, is capable of both action and feeling, 'Laertes (qui aurait plutôt mérité, j'y songe, hélas! trop tard, d'être le héros de cette narration)'.[7] Kate, who has not witnessed the murder, assumes that he has committed suicide for his 'true love', Ophelia. And this is the false story that she proclaims to the world. 'Un Hamlet de moins; la race n'en est pas perdue, qu'on se le dise!', is the narrator's final comment. And the reader remembers Horatio's very different words:

> Now cracks a noble heart.—Goodnight, sweet prince;
> And flights of angels sing thee to thy rest!

What we are given is a systematic refusal of the heroic tone, a reduction of Shakespeare's tragic inevitability to a burlesque sequence of chance incidents. The *moralité* abounds in parodic echoes of Hamlet's memorable pronouncements: 'To be or not to be'; 'Words, words, words'; 'Frailty, thy name is woman!' ('Stabilité! ton nom est Femme').[8] They resound nostalgically in the wings of Laforgue's brutally diminished tale. And this is not all. Laforgue explores the inner motivations of Hamlet's psychology relentlessly. Frustration leads to sadism, and, in a parody of Flaubert's massacre of animals in 'La Légende de Saint Julien l'Hospitalier' (which draws on the theme of Hamlet's cruelty to both Ophelia and his mother), Laforgue clearly intends to alienate the reader from his hero. If the eponymous hero of Musset's *Lorenzaccio*, moulded as he was on the figure of Hamlet, offered a devalued Hamlet

[6] Known as the Saxo Grammaticus version. Cf. W. Hansen, *Saxo Grammaticus and the Life of Hamlet* (Lincoln, Nebr., 1983). [7] *ML* 45. [8] Ibid. 9.

to the early nineteenth century—'le seul Hamlet que nous méritions', Musset might have said—he at least had the stature of the anti-hero. Laforgue's hero cannot lay claim even to this. In the field of action he is quite simply insignificant. A passive victim rather than a 'just' murderer, he represents the ultimate in inaction.

That the *moralité* is highly effective parody is not in doubt. As Laforgue changes the plot of the play, so he reverses the importance of the play's elements. The political and revenge themes are reduced. The psychological interest of Hamlet's horror at his mother's infidelity is largely removed by the fact that she is not his actual mother. Her behaviour is significant only in that it is symptomatic of female behaviour in general, and thus dictates Hamlet's attitude towards Ophelia. The love theme, secondary in Shakespeare's play, becomes primary. But even this is no more than part of a more global preoccupation, that of registering the modern consciousness, which is projected in a narrative visibly concerned with the disparity of levels that it offers. That the disenchanted voice should be that of the artist is highly appropriate in a century during which the artist had progressively embraced this role. Hamlet as artist is thus a theme discovered in the margins of Shakespeare's text and given central importance.

In the *moralité*, therefore, ironic 'dédoublement' reveals similarities beneath discrepancies, and vice versa. In the juxtaposition of epigraph and poem there is a parallel relationship. Shakespeare's play is valuable to the Laforgue of the *Fleurs* in two ways. First, to stress the distance between Laforgue's situation and the transcendence offered by tragedy. The characters of Shakespeare's play inhabit ultimately a world of true values. The cathartic experience of the play's ending restores harmony. Hamlet and Ophelia have become 'noble' victims of fate. Secondly, to stress similarity, because this ambiguous and undeveloped relationship is surprisingly modern in its elusiveness.[9] We see it only in fragmentary glimpses, and are forced to decode it at every stage. Hamlet's brittle wit, Ophelia's faltering replies reveal levels quite different from the surface words they utter. Each character inhabits a closed universe, and each is incapable of understanding the other. These are themes that preoccupied Laforgue, and which were later to displace more conventional ones in the depiction of love by a writer such as Proust. Ophelia is a victim, a pawn, both of her father and of the corrupt society in which she lives,

[9] 'The attitude of Hamlet towards Ophelia is without doubt the greatest of all the puzzles in the play,' Dover Wilson wrote in 1935 (J. Dover Wilson, *What Happens in Hamlet* (Cambridge, 1974), 101).

and parallels between her and the young bourgeois girls who practised the 'éternelles ritournelles' that figure so insistently in Laforgue's verse are quite clearly intended to be drawn. The value to Laforgue of the Hamlet scenario resides, then, in the modernity of the relationship—the sense of incomprehension and complexity that it conveys—but, at the same time, in the elevation of tragedy, because, of course, the 'modern' view does not consist only of the confusion of meanings of the present: it contains also the nostalgia for the 'true values' offered by tragedy. ('The lack of a destiny is', as Jonathan Culler observes, 'the modern form of the tragic.'[10]) As one reads the *Fleurs* one cannot escape the reflection cast on the poems by the fragments of dialogue from *Hamlet*. The image of the Hamlet/Ophelia relationship—with its similarity and its difference—is hauntingly present. Thus, by the parallels and the contrasts that it offers, the fictional creation of an author long since dead acts as an effective intertext for a young writer obsessed by a sense of the ephemeral.

'UN LIVRE ENFIN DE BONNE FOI': LOVE AND ART

In the final poem of the *Lunes* Laforgue had promised his reader 'un livre enfin de bonne foi', but, as we have seen, this protestation of future honesty was undercut by irony. 'Avertissement', the introductory poem of the *Fleurs*, however, sets about fulfilling the promise. It is similar in function to *Au lecteur*, with which *Les Fleurs du mal* opens, and it is quite different, as one would expect, in tone and character. Laforgue, who parodied Baudelaire's title in his own, no doubt intended both the parallel and the parody. On the manuscript he wrote 'préface sur la couverture même', and above this: 'aux français de demain'. The poem's title plays on the word's double meaning; this is a warning as well as an introduction. Like the drunken Spartan slave, the poet will caution future generations, not against the excesses of alcohol, but against that modern disability, the rootless personality. In a striking metaphor he compares himself to a weathervane: 'Bonne girouette aux trent' six saisons' (the fanciful thirty-six was originally four).[11] The inspiration of the poem was drawn from Laforgue's visit to Elsinore, 'le pays de Hamlet',[12] on 1 January 1886. Two areas of discussion are

[10] J. Culler, *Flaubert: The Uses of Uncertainty* (Ithaca, NY, 1974), 145.

[11] The image is used also in his correspondence: 'n'être plus qu'une ingénue girouette aux quatre saisons' (*LA* 165).

[12] Ibid. 143.

established: the emotions and art; the paralysis of feeling—the 'blocus sentimental' of the first poem in the *Derniers Vers*—and creative self-doubt. Laforgue's narrator is as self-deprecating as are Beckett's characters: 'Signifier? Nous, signifier! (Rire bref.) Ah elle est bonne!'[13] Verlaine, whose Gaspard Hauser lament[14] might be compared with this, is sentimental by contrast. The poem is both very personal and disconcertingly objective. It sets out the state of affairs with what appears to be a shocking refusal of emotion.[15] The metre it employs for this aim is the nine-syllable *impair* line, overrunning the lyrical octosyllable, and stopping short of the more dignified decasyllable.

Mon père (un dur par timidité)
Est mort avec un profil sévère;
J'avais presque pas connu ma mère,
Et donc vers vingt ans je suis resté.

Alors, j'ai fait d'la littérature;
Mais le Démon de la Vérité
Sifflotait tout l'temps à mes côtés:
'Pauvre! as-tu fini tes écritures...'

Or, pas le cœur de me marier,
Étant, moi, au fond, trop méprisable!
Et elles, pas assez intraitables!!
Mais tout l'temps là à s'extasier!...

C'est pourquoi je vivotte, vivotte,
Bonne girouette aux trent' six saisons,
Trop nombreux pour dire oui ou non...
—Jeunes gens! que je vous serv' d'Ilote!

In its rigorous detachment, 'Avertissement' may be seen as a successful example of the goodwill of the volume's title. The tameness of 'bonne volonté', as against evil, and the casualness of 'des', as against the dogmatism of Baudelaire's 'les', are precisely what is meant. These are a few disparate flowers, and goodwill is all that the modern anti-hero can hope to possess. For what is a poet but, in Mallarmé's words (though Laforgue could not have known them), 'un instrument qui résonne sous les doigts des diverses sensations'?[16] The volume

[13] S. Beckett, *Fin de partie* (n.p. 1957).

[14] P. Verlaine, *Sagesse* III. iv.

[15] Laforgue reacted very emotionally to what he interpreted as his father's possible disapproval, as the following passage from a letter to his sister Marie shows: 'Comment ai-je le courage de penser à autre chose qu'à ma douleur de fils, à la douleur de n'avoir pas revu papa, de ne pas savoir ce qu'il pensait de moi en s'en allant. . . . Est-il parti espérant en moi?' (*AH* 705.)

[16] Stéphane Mallarmé, *Correspondance*, i. ed. H. Mondor (Paris, 1959), 151.

represents a significant new direction, a move back to centre, one might say, after the 'mince plaquette' that was the *Lunes* and the *Quelques Complaintes de la vie*, which grew to fifty because publication was irritatingly delayed. It is an exploratory volume. The large number of variants that survive, and which will be examined in the next chapter, reveal just how much Laforgue was feeling his way. Written between April and June 1886 in a burst of intense poetic activity,[17] it was abandoned when Laforgue moved into free verse. Michael Collie writes of the *Fleurs*: 'What we have here as *Des Fleurs de bonne volonté* is no more than a writer's notebook, "un répertoire pour de nouveaux poèmes"',[18] and it is true that this is what it *became*, since, as we shall also see in the following chapter, Laforgue dipped into a range of poems from the *Fleurs* and quoted them (with significant alterations) in the *Derniers Vers*. There were, however, two stages: one in which the volume represented an exciting and important new advance from the *Lunes*, and another, following soon after, and in the wake of his revolutionary formal breakthrough, in which the volume was converted into a 'writer's notebook'. This is proved, if proof is needed, by a manuscript 'table de matières', giving the titles of fifty-six poems, as well as by a title-page complete with the name of his proposed publisher (Léon Vanier) and publication place and date: Paris 1886.[19]

Unlike the *Lunes*, which consisted of two main types of poem, those describing the moon and those centring on the Pierrot figure, the *Fleurs* are characterized by extreme formal variety. Narrative, dramatic, analytical, aphoristic, and parodic poems are all present. Whatever the mode, however, thematically they all explore the same central area: the instability and essential isolation of the self, and the desire for the commitment and reassurance of love—two sides, obviously, of the same coin. A wide range of moods with regard to love is expressed, from a sense of profound inadequacy and hopelessness, to the registering of moments of hope.

The order of the poems follows the interior rhythms of the mind rather than any rational structure. There is some use of parallelism, but more frequent use of juxtaposition of contrasting moods or tones in a

17 'Eh! bien, j'ai maintenant 35! (trente-cinq) pièces (et plusieurs point courtes) de ce volume, au net. Est-ce que ça ne t'effraie pas pour moi?' (*LA* 183.)

18 M. Collie, *Jules Laforgue* (London, 1977), 59.

19 Now in the Bibliothèque Jacques Doucet, Paris. E. Dujardin confirms this intention. 'Laforgue cessa de travailler à *Des Fleurs de bonne volonté* en 1886. A cette époque, bien que la rédaction n'en fût pas définitive, il eut l'intention de les publier chez l'éditeur Léon Vanier.' (Dujardin, *Poésies complètes de Jules Laforgue* (Paris, 1894), p. v.)

technique of counterpoint rather than of symmetry. The only repeated pattern is that of the poet's 'bonne volonté' meeting obstacles that are metaphysical, personal, and sexual in nature. If the anguish of the metaphysical theme is constantly present, it is perhaps more insistently so in the early part of the volume; while the projection towards a future of possible, if doubtful, fulfilment in love is slightly more prominent, as one might expect, in the second half of the volume. 'Dimanches' poems are interspersed throughout the volume, as are 'Petites Misères'—without regard to seasonal sequence—and, if these have a slight stabilizing effect on the structure of the whole, nevertheless the volume, which ends with two of its most whimsical compositions, is ultimately, as is appropriate, both tentative and inconclusive.

The 'modern' sensibility, unsupported by religious faith, conscious of isolation and the fragmentation of the personality, contemplating the impossibility—or, at times, possibility—of satisfaction in love: this is the subject-matter that is being insistently investigated. Jacques Rivière considered Laforgue's area a narrow one: 'je lui reproche de réduire perpétuellement toute l'histoire, toute la littérature et toute la fantaisie à sa vision. Car enfin le malentendu entre les sexes, c'est très joli mais c'est connu et peu intéressant.'[20] One could counter this objection by saying that what Laforgue does when he describes with extraordinary accuracy his 'inutile cœur d'adolescent'[21] is to illuminate a moment when life is still viewed from the sidelines. 'Only the young can be alone freely',[22] wrote Philip Larkin, and it could be argued that Laforgue drew richly from that period of youthful isolation when the personality has not yet embarked on the journey of adaptation, compromise, *Bildung*, and when the interrogative stance towards life is, therefore, found in its purest form, combining freedom and fixation, mobility and obsession. Rather as Rimbaud had seemed to exhaust the poetic possibilities of a certain kind of youthful rebellion, so Laforgue focused on a certain kind of adolescent yearning. 'Laforgue, tant sa puberté était riche, s'en alimente', wrote Benjamin Crémieux, 's'en torture et ne consent pas à l'épuiser.'[23]

But insipidity could well have resulted from such inevitably limited self-exploration were it not for two things: the continuous metaphysical backing to the poems, and Laforgue's quite exceptionally rigorous

[20] Jacques Rivière and Alain-Fournier, *Correspondance*, i (Paris, 1926), 232.

[21] XL, 'Petites Misères d'automne'.

[22] P. Larkin, 'Vers de société', *High Windows*.

[23] B. Crémieux, 'Compte-rendu des *Dragées*', *NRF*, Mar. 1921.

lucidity. He is the intellectual who brings his mental powers to bear on the emotional situation. T. S. Eliot pinpointed the central problem when he described his dilemma as the 'somewhat despairing recognition that what he knows does not permit what he feels, and that what he feels contradicts what he knows'.[24] 'Despairing recognition' is accurate, because Laforgue's intellect is intent on observing and registering; it does not follow the paths of rebellion or attempted transcendence. Like Beckett's work of this title, his poems describe 'Comment c'est'—obviously at a certain time, with a certain personality, in a certain situation—but, across the particular, the universal is glimpsed. If one values fidelity in the observation of an object, however apparently insignificant, then Laforgue's sensitive listening to the patterns of the emotions as they meet and interact with the conscious intelligence will seem not only justified, but necessary and, in some ways, parallel to more sustained later exercises of a similar kind to be found in a writer such as Proust.

In this task he drew inevitably on his intimate personal experience, and it has been suggested that it was in part because this volume reflected so closely his misgivings and internal conflicts concerning his love for Leah Lee that he did not publish it in the end. It is, according to Reboul, 'le journal intime d'un jeune homme qui hésite devant un engagement décisif'.[25] But much of the hesitation is self-directed, and the longing, sometimes expressed with direct simplicity, is intense:

> Sucer la chair d'un cœur élu,
> Adorer de souffrants organes,
> Etre deux avant qu'on se fane!
> Ne serai-je qu'un monomane
>   Dissolu
> Par ses travaux de décadent et de reclus?
> ('Célibat, célibat, tout n'est que célibat')

Nevertheless, this longing is clearly only one element in a complex configuration. This is true also of other extreme statements, the urgent plea for fraternal relations between the sexes, for example—unusual at this time, and noted for this reason by Simone de Beauvoir in *Le Deuxième Sexe*.

[24] M. Hannoosh, 'Metaphysicality and Belief: Eliot on Laforgue', *Comparative Literature*, Autumn 1987.

[25] P. Reboul, *Laforgue* (Paris, 1960), 151.

Je te crierai: 'Nous sommes frères!
Alors, vêts-toi à ma manière,
Ma manière ne trompe pas;
Et perds ce dandinement louche
D'animal lesté de ses couches,
Et galopons pas les haras!'
('Dimanches XXXIV')

The vehemence here is genuine, but the statement emerges as nothing more than a rhetorical outburst—an expression of feeling, mocked by its extremity and the animal imagery that Laforgue uses. The poet is denied a privileged status above the fray or even as a mediator between opposites. These dramatized monologues and snatches of narrative offer us instead disturbing shafts of partial illumination. If a Baudelairean dualism still roots the texts thematically (the polarities of antagonism and concern, domination and worship, cruelty and tenderness, for example), our impression is one of dispersal even more than of division. We are given no dominant image of the poet. Laforgue could not even have said with Rimbaud: 'Je me suis reconnu poète.'[26] He is closer to Mallarmé, who described himself as 'multiple, impersonnel', [27] or to Keats when he wrote:

> A Poet is the most unpoetical of any thing in existence; because he has no Identity—he is continually informing and filling some other Body—the Sun, the Moon, the Sea and Men and Women who are creatures of impulse are poetical and have about them an unchangeable attribute—the poet has none; no identity—he is certainly the most unpoetical of all God's Creatures.[28]

'DU TÊTE-À-TÊTE AVEC SOI': NARRATIVE AND DRAMATIZATION

Laforgue's last poems, the *Derniers Vers*, employ a dramatized interior monologue that has some of the characteristics of the twentieth-century stream of consciousness. For its effect, however, this monologue frequently depends on a further narrative element that roots the meditation in a 'real' event and lends it a necessary past. Such narrative 'cores' are in fact frequently lifted from the *Fleurs* and embedded in the symphonic structures of the *Derniers Vers*. In the earlier volume Laforgue can be

26 'Lettre du voyant' to Georges Izambard, May 1871, Arthur Rimbaud, *Œuvres complètes* (Paris, 1951), 252.
27 Mallarmé, *Correspondance*, i. 407.
28 J. Keats, *Letters*, ed. M. Buxton Forman (Oxford, 1935), 228.

seen to be experimenting with narrative in a new way. Narrative in the *Complaintes* had been either borrowed from folk-song and myth, or extremely elliptical. This tended to give a representative quality to its 'characters'. In the *Fleurs* the particularized human figure is developed. There can be no doubt that the composition of poems like the following must have played an important part in Laforgue's poetic development, even if he ultimately moved away from narratives of this kind. The banal miniature tragedy of 'Dimanches XVI' has the colouring of reality, which was not the case with earlier successful narratives like 'Complainte du roi de Thulé', and is still not the case in the more symbolic narratives of the *Fleurs*.

> HAMLET: Have you a daughter?
> POLONIUS: I have, my lord.
> HAMLET: Let her not walk i' the sun;
> conception is a blessing; but not [*sic*]
> as your daughter may conceive.

Le ciel pleut sans but, sans que rien l'émeuve,
Il pleut, il pleut, bergère! sur le fleuve. . .

Le fleuve a son repos dominical;
Pas un chaland, en amont, en aval.

Les Vêpres carillonnent sur la ville,
Les berges sont désertes, sans idylles.

Passe un pensionnat (ô pauvres chairs!)
Plusieurs ont déjà leurs manchons d'hiver.

Une qui n'a ni manchon, ni fourrures
Fait, tout en gris, une pauvre figure.

Et la voilà qui s'échappe des rangs,
Et court! ô mon Dieu, qu'est-ce qu'il lui prend?

Et elle va se jeter dans le fleuve,
Pas un batelier, pas un chien Terr' Neuve.

Le crépuscule vient; le petit port
Allume ses feux. (Ah! connu, l'décor!)

La pluie continue à mouiller le fleuve,
Le ciel pleut sans but, sans que rien l'émeuve.

In his earlier version of this poem Laforgue had introduced elements that indicated directly the cause of the girl's suicide. In his final version he is more reticent, and allows the epigraph from *Hamlet* to make the point. Variants show also that the poem's skilfully handled emotional restraint was not Laforgue's first intention.[29] In his final version he achieves a balance of involvement and detachment, pathos and wit. His stance is simultaneously sensitive to the tragedy and aware of its insignificance in the larger scale of human affairs. The girls from the convent are not the expected 'âmes', but are 'pauvres chairs', punning on 'chères'.[30] The river, rather than any of the characters, is enjoying its Sunday rest. The legendary 'chien Terr' Neuve', which might have saved the girl had it been present, is deprived of a syllable. The last line repeats the first in ballad style, and a well-known folk-song—'Il pleut, il pleut, bergère'—has been offered as an intertext in the second line. Thus, a totally unpretentious poem succeeds in suggesting the cruelty of life where Laforgue's early Hugolian attempts had failed.

This ability to convey intense emotion in brief, often disjointed, narratives is found in several poems of the *Fleurs*, which evoke a fragile and compelling female figure in a new way:

(Une, pourtant, je me rappelle,
Aux yeux grandioses
Comme des roses,
Et puis si belle!. . .
Sans nulle pose.

Une voix me criait: 'C'est elle! Je le sens;
Et puis, elle te trouve si intéressant!'
—Ah! que n'ai-je prêté l'oreille à ses accents!. . .)
('Dimanches XXXVI')

Une maigre qui me parlait,
Les yeux hallucinés de Gloires virginales,
De rendre l'âme, sans scandale,
Dans un flacon de sels anglais. . .
('Dimanches XXXVIII')

Je me souviens,—dis, rêvé ce bal blanc?
Une, en robe rose et les joues en feu,
M'a tout ce soir-là dévoré des yeux,
Des yeux impérieux et puis dolents,
(Je vous demande un peu!)
('Petites Misères d'automne XL')

29 These will be illustrated in the following chapter, in which variants are discussed.

30 That the pun was conscious is illustrated by a fragment—probably from the projected novel *Saison*—which ends: 'A moi, ma chère . . .—chair . . .' (*FV* 189).

These 'caillots de souvenir',[31] dense, emotionally charged memories, are a valuable element in the orchestration of the *Derniers Vers*. In the following chapter we shall examine the use made of two poems from the *Fleurs* ('Arabesques de malheur LII' and 'Dimanches XLIV') in the *Derniers Vers* poems that incorporate them. From the quoted fragments, as from 'Dimanches XVI', we can already appreciate the importance of irony in the Laforguian narrative. The phrases 'Et puis elle te trouve si intéressant', 'Dans un flacon de sels anglais', and '(Je vous demande un peu!)', by restoring the narrative equilibrium, make it possible for the reader to respond, without too much scepticism, to the emotional appeal that is being made.

The dramatic element in the *Fleurs* is still more striking, and, indeed, Laforgue's poetry was intrinsically dramatic from the start, as we have seen. His first sustained work was a verse play (*Tessa*, 1877). His first volume of poems, influenced by the grand cosmic rhetoric of his predecessors, set up a dramatic confrontation between man and the universe, personifying the latter. The *Complaintes* originated from spoken or sung models, and brought to life a multiplicity of voices. In the *locutions* of his Pierrots in the *Lunes* an internal dialogue with the self was developed. This is the aspect that the *Fleurs* essentially took further, as can be seen in *Le Concile féerique*, a fanciful drama published in *La Vogue* in July 1886, which is based on a range of poems from the *Fleurs*. Here, in a highly inventive manner, Laforgue uses three poems of interior monologue, breaking them up into the speeches of his dramatis personae—le Monsieur, la Dame, le Chœur, and an Echo—and combines them with two poems that are already speeches, that of the cynical Monsieur ('Esthétique XLII'), and that of the Dame, who represents the eternal feminine ('Notre petite Compagne XLV'). These witty, stylized poems, flippant in tone, musical in their regular octosyllabic metre, strongly reminiscent of the speeches of the Pierrots in the *Lunes*, form an effective contrast to the more complex substance drawn from the 'Petites Misères' of July and August, and from 'Le Bon Apôtre XXII'. (The interior monologue of these particular poems probably gains by being set out as the speeches of various characters, as this provides a simplification of texts that appear at points fruitlessly dense.) Intended for the stage, probably rather in the manner of Huysmans's and Hennique's *Pierrot sceptique*, this collection of poetic speeches was not in fact performed until 1893, at Paul Fort's Théâtre d'art. Its

[31] 'Légende VIII', *Derniers Vers*.

publication in 1886 must, however, have had some influence on Laforgue's decision not to publish the *Fleurs*. That he was prepared to remould his poems to form a play illustrates the closeness of his poetry to this art-form.

The most important advances in techniques of dramatization can best be seen, however, in other poems: in 'Avant-dernier Mot', for example:

L'Espace?
—Mon Cœur
Y meurt
Sans traces. . .

En vérité, du haut des terrasses,
Tout est bien sans cœur.

La Femme?
—J'en sors,
La mort
Dans l'âme. . .

En vérité, mieux ensemble on pâme
Moins on est d'accord.

Le Rêve?
—C'est bon
Quand on
L'achève. . .

En vérité, la Vie est bien brève,
Le Rêve bien long.

Que faire
Alors
Du corps
Qu'on gère?

En vérité, ô mes ans, que faire
De ce riche corps?

Ceci,
Cela,
Par-ci
Par-là. . .

En vérité, en vérité, voilà.
Et pour le reste, que Tout m'ait en sa merci.

The form of the poem clearly builds on the counterpoint technique developed in the *Complaintes*: a third voice offers comments on the central debate, which is expressed in the poem's disyllabic quatrains.

The third voice was a late addition (the two earliest manuscripts do not contain it). Always beginning 'En vérité', it offers a rational—and futile—comment on the situation:

> En vérité, la Vie est bien brève,
> Le Rêve bien long.

The fact that the comment is a platitude does not exclude wit, a situation exploited by twentieth-century Absurdist theatre. Here it is seen in the fact that 'Rêve', capitalized, replaces the expected 'Art', and underlines the brevity and apparent frivolity of this particular fragment of art. At a later point in the poem, in a technique reminiscent of that employed by Verlaine in his similarly inconclusive poem 'Triste, triste était mon âme',[32] question is 'answered' by question: 'En vérité, ô mes ans, que faire | De ce riche corps?' In Verlaine's poem the 'tête-à-tête avec soi' (the phrase Laforgue used to describe Verlaine's poetry[33]) is between the heart and soul, and the soul is quite incapable of enlightening the heart. Laforgue's poem is not about inexplicable emotional turmoil, but about the human condition in a broader sense. It ends with the evocation of the void with which it had opened, illustrated by a breakdown of the poem's regular metre. The final couplet, consisting of a decasyllable followed by an alexandrine, has no antecedent. Until this couplet, disyllabic quatrains have alternated regularly with imparisyllabic couplets (9, 5). 'En vérité' is repeated now for the first time: 'En vérité, en vérité, voilà.' There is quite literally no more to be said, and the language perfectly expresses this state of affairs. Laforgue concludes with a prayer addressed not to God, but to a capitalized 'everything', and written in the most intentionally unrhythmical of alexandrines. This breakdown of syllabic pattern within a conventional verse structure illustrates the degree of experimentation that Laforgue was prepared to allow himself, and anticipates, of course, his move into free verse. A kind of Baudelairean 'soubresaut de la conscience'[34] (which, for Baudelaire, necessitated the use of prose), it makes Laforgue's concern with the equivalence of form and meaning perfectly clear. Throughout the poem there is subtle play with formal devices—with repetition to accentuate or slightly alter sense, with phonetic patterning, with suspension points to extend disyllabic lines, or, as we

32 P. Verlaine, 'Ariettes oubliées VII', *Romances sans paroles.*

33 *LA* 54. In the *Agenda* of 1883 Laforgue wrote of Verlaine: 'Quel vrai poète—C'est bien celui dont je me rapproche le plus.' (*AH* 886.)

34 Charles Baudelaire, dedication of *Le Spleen de Paris* to Arsène Houssaye.

have seen, with humorous capitalization—devices that, like the final break with the anticipated pattern, have been used self-consciously, to mark shifts of stance. In this elusive poem the reader must attempt to discover whatever meaning he can from the 'space' created between its three voices, dramatic projections designed to challenge and disappoint our desire for understanding and synthesis.

If we take more obviously psychological poems, such as 'Mettons le doigt sur la plaie' and 'Dimanches XXXVI', with less clearly defined 'voices', we can observe a similar process at work:

Que le pur du bonheur m'est bien si je l'escompte!. . .
Ou ne le cueille qu'en refrains de souvenance!. . .
O rêve, ou jamais plus! Et fol je me balance
Au-dessus du Présent en Ariel qui a honte.

Mais, le cru, quotidien, et trop voyant Présent!
Et qui vous met au pied du mur, et qui vous dit:
'A l'instant, ou bonsoir!' et ne fait pas crédit,
Et m'étourdit le cœur de ses airs suffisants!

Tout vibrant de passé, tout pâle d'espérance,
Je fais signe au Présent: 'Oh! sois plus diaphane?'
Mais il me bat la charge et mine mes organes!
Puis, le bateau parti, j'ulule: 'oh! recommence. . .'

Et lui seul est bien vrai!—mais je me mords la main
Plutôt (je suis trop jeune. . . ou, trop agonisant. . .)
Ah! rien qu'un pont entre Mon Cœur et le Présent!
O lourd Passé, combien ai-je encor de demains?. . .

O cœur aride
Mais sempiterne,
O ma citerne
Des Danaïdes!. . .

'Mettons le doigt sur la plaie' is a poem about the difficulty of living in the present, and thus concerns man's relationship with time. No single image dominates. The narrator, an insubstantial Ariel figure, is seen in many stances: forced to the wall in a military image of shooting, refused credit in a commercial image, nostalgically picking the flowers of past happiness, missing the boat of lived experience. The abstract subject-matter has been given concrete form not by a dominating image, nor by a 'real' landscape, but by a swiftly moving series of poetic personae. The poem gives a sharp impression of the problem of living in the present ('le cru, quotidien, et trop voyant Présent') for the sensibility that responds more naturally to past and future ('Tout vibrant de passé, tout pâle d'espérance'). But, no matter how vibrant its echoes in the present, Laforgue refuses to endow the past with the privilege of meaning. We have instead a plea for participation in present experience. However, since the narrator has shown all too clearly everything that makes the present unacceptable, the problem and its insolubility are simultaneously offered to the reader, the desire and its subversion. The poignancy of the poem depends on this ambiguity, and, since this is true of much of Laforgue's poetry, his self-analysis can be said to lead naturally and unavoidably to inconclusive self-questioning.

The final quatrain stands apart from the rest of the poem in function as well as in style, and laments the situation that has been so graphically described earlier:

O cœur aride
Mais sempiterne,
O ma citerne
Des Danaïdes!...

The comment of a third voice in the debate introduces a new angle of vision, and appeals to the reader's judgement and sense of values. If in this case it provides an acceptable realistic assessment, it again cannot offer a way forward. Its irony is of the kind to which Flaubert referred when he wrote: 'L'ironie n'enlève rien au pathétique: elle l'outre au contraire.'[35] The same is true of 'Dimanches XXXVI':

J'aurai passé ma vie à faillir m'embarquer
Dans de bien funestes histoires,
Pour l'amour de mon cœur de Gloire!...
—Oh! qu'ils sont chers les trains manqués
Où j'ai passé ma vie à faillir m'embarquer!...

Mon cœur est vieux d'un tas de lettres déchirées,
O Répertoire en un cercueil
Dont la Poste porte le deuil!...
—Oh! ces veilles d'échauffourées
Où mon cœur s'entraînait par lettres déchirées!...

Tout n'est pas dit encor, et mon sort est bien vert.
O Poste, automatique Poste,
O yeux passants fous d'holocaustes,
Oh! qu'ils sont là, vos airs ouverts!...
Oh! comme vous guettez mon destin encor vert!

[35] Gustave Flaubert, *Correspondance*, ii (Paris, 1980), 172.

(Une, pourtant, je me rappelle,
Aux yeux grandioses
Comme des roses,
Et puis si belle!. . .
Sans nulle pose.

Une voix me criait: 'C'est elle! Je le sens;
Et puis, elle te trouve si intéressant!'
—Ah! que n'ai-je prêté l'oreille à ses accents!. . .)

Here the poet's instability is effectively suggested by bewildering projections to a romantically viewed past and a compelling but uncertain future. A vivid evocation of 'the mind simultaneously possessed by diverse and even conflicting feelings'[36] that it is unable to order or dominate, the poem shows how contrasting stances complement one another and even meet. So illogical—and yet so natural—are the transitions from one set of impulses to another that, at each stage of the poem, the reader is invited to respond ironically. But he also responds emotionally, in part because of the linguistic accuracy with which these transitions are rendered. The apostrophized eyes ('O yeux') are reduced to an incidental part of the descriptive background ('Aux yeux grandioses'). The hypothetical 'J'aurai passé me vie' becomes hard fact: 'j'ai passé ma vie'. 'Sort' is transmuted to the more threatening 'destin'; 'bien vert' to the wistful 'encor vert'. Every exclamatory 'Oh', 'O', or 'Ah' is calculated and appropriate to its precise context.[37] Even in this short poem we can see how the dramatic technique effectively brings to life the tumult of the diverse elements of the consciousness, and systematically refuses to do more than this. The poem combines the dramatic monologue that we have been considering with a final, highly effective narrative backdrop—set out entirely within brackets, in parenthesis, as it were; that is to say, it offers the reader in miniature the ingredients of some of the most successful of the *Derniers Vers*.

The poem's complexity is echoed in the liberties that Laforgue takes with its versification. As in 'Avant-dernier Mot', a pattern is set up that is then broken, dissolving like the initially stable stage-set in an Ionesco play. The three opening stanzas regularly enclose three octosyllabic

[36] M. Turnell, 'The Poetry of Jules Laforgue', *Scrutiny*, 5, Sept. 1936.

[37] Clive Scott develops this point in *A Question of Syllables* (Cambridge, 1986), 166–80. The ironically capitalized 'Poste' masks a biographical irony, if we are to believe the very plausible story told by E. Dujardin. Laforgue did in fact retrieve from the Imperial Mail a letter containing a proposal to Leah Lee (G. Jean-Aubry MSS.).

lines in alexandrines (12, 8, 8, 8, 12). This regularity is destroyed by the two final stanzas, set within brackets, which represent an afterthought that is also an interruption, a vivid intrusion of memory into the present. As in Beckett's *La Dernière Bande*, the memory that disrupts the meditation is the live element in the piece, signalled here by a change of metre and an abandonment of regular form. An octosyllable is followed by a line of five syllables, after which come three tetrasyllabic lines. But form is not altogether abandoned, because three of these rhythmical, if unequal, lines are linked (and softened) by feminine rhyme. The last of them ('sans nulle pose') drives home the contrast between the apprehension of the 'real' and the realization of the illusory, which has preceded them and which now follows them in three regular and strongly linear alexandrines, expressing vain regrets underscored by irony. As in Laforgue's later free verse, technical transgressions have a precise semantic justification. Infringements of the norm are enjoyable curiosities,[38] but they are also a cause of subtlety: they may lead to the evocation of the elusive 'insaisissable'[39] that Laforgue, like his contemporaries, sought to convey.

Ironies can be explored within this subversion of regularity. Should one, using the technique of synaeresis on 'grandiose' (mocking the word's elevated meaning), bring this five-syllable line down to four, making it identical with the three lines that follow it? Should one, then, pursuing semblances of regularity, read the stanza's opening octosyllable as the two tetrasyllables into which it teasingly falls? One thing is certain: the lines of enquiry suggested by the uncertain status of the form that Laforgue has chosen to employ here—as in a number of other poems in this volume[40]—match the ambiguity and circularity of the complex mental processes that represent his thematic base.

Letters to his friends bear out the impression one gains from the poetry of a compelling interest both in the technicalities of versification and in its liberation. His sense of closeness to Verlaine—since the two poets inhabit very different intellectual worlds—must have depended to

[38] 'En attendant c'est plus de rêve, de choses tenues, de curiosités', Laforgue advised Kahn (*LA* 68).

[39] 'On y perd en insaisissable', Laforgue wrote of Kahn's use of the alexandrine (ibid. 65).

[40] Other poems in which the established syllabic regularity disintegrates are: 'Dimanches XXXVIII', 'Dimanches XLIV'. In some poems the final stanza only is in a different metre, as in III and LIV. In the latter the metre that is broken is extremely intricate. Stanzas 1 and 3 are as follows: 10, 12, 10, 3, 12, 10; stanzas 2 and 4 balance 1 and 3 as follows: 12, 10, 12, 3, 10, 12; stanza 5 breaks the rhythm altogether: 8, 8, 8, 8, 5.

a considerable degree on this recognition of a parallel poetic endeavour, with Verlaine's partial liberation of metre being used also to portray indeterminate states of consciousness and to capture modern 'nuances' of sensation and of sexual innuendo.[41] Whenever Laforgue gives his friends advice on particular poems they have sent him—and he does this especially with Mme Mültzer and with Kahn—he urges them towards novelty and experimentation. One area is rhyme. Already in 1882 he expressed horror to Henry at the banality of Mme Mültzer's rhyme schemes: 'Par exemple ce sonnet 'Les Elus' sur rimes données!'[42] When he addressed her personally, he was more polite: 'Vous verrez qu'il ne faut pas rimer *éperdues*, *étendues*, *fantastique*, *extatique*, *fleurs*, *cœurs*.'[43] On the other hand, an outlandish rhyme of Kahn's causes evident pleasure ('La rime tessons m'a fait jubiler comme tu penses'[44]), and he writes to Henry: 'Kahn m'a envoyé une belle pièce singulière d'exécution dans le sans-gêne de ses rimes.'[45] Another area is syllabic length. He criticizes Kahn's fondness for the alexandrine, which seems to him ill-suited to his subject-matter, that is, to the 'pièces si balbutiées de langue et si infinies de décor'[46] of the early *Palais nomades*. The alexandrine lacks the flexibility that contrasting syllabic lengths would supply: 'Et surtout impossible de s'y livrer comme avec des strophes à part, à cette distribution en staccato et en menus enroulements et déroulements fugués qui est devenue pour moi un besoin, une envie de femme enceinte dans les vers de rêve.'[47]

Laforgue adapts his choice of metre to his subject-matter. He will still use the alexandrine, but frequently at this date it is 'with a difference', often, as in 'Petite Prière sans prétentions' and 'Le Bon Apôtre', for its potential for parodic subversion. The octosyllable is used for its melodic lightness, but Laforgue clearly enjoys also the ambivalence that results from the use of this metre for intellectual or scientific material. 'Cas rédhibitoire' is a prime example. The intricately patterned stanzas found in many *Complaintes* are developed still further here, with an extreme sensitivity to the different rhythms created by varying syllabic

[41] Laforgue's references to Verlaine bear this out, for example: '*Sagesse* où il y a des choses épatantes pour moi, de la vraie poésie, des vagissements, des balbutiements dans une langue inconsciente ayant tout juste le souci de rimer.' (*AH* 845.)

[42] Ibid. 732. [43] Ibid. 764. [44] *LA* 69.

[45] *AH* 803. Laforgue discussed such details of prosody with Henry because, scientist though he was, he was both fascinated by the principles of aesthetics and was an aspiring poet himself. The first number of *La Vogue* published a prose poem by him, 'Vision', which disappointed Laforgue (*LA* 176). [46] *LA* 65. [47] Ibid.

lengths.[48] The point is not without importance, because the success of Laforgue's later free verse will depend in part on his response to these resonances. By subtle sleights of hand, he obliges us to take note both of traditional boundaries and of transgressions of these boundaries, and, consequently, of the poem itself both as a completed artefact and as a series of mobile creative possibilities. He plays, as on a musical instrument, on our anticipated reactions of surprise, shock, or amusement. There is also clear enjoyment of the virtuosity that such constructions represent. Sometimes, now, they are set up, as we have seen, for the almost equal pleasure of their demolition. This is accomplished with the 'sans-gêne' that Laforgue occasionally found in Kahn and always admired. It is, as we know, a contrived 'sans-gêne'. He is still the Pierrot as regards the technical exploitation of artifice, and every successful poem in the *Fleurs* is at one level a theatrical performance that delights in the writer/performer's manipulation of his audience.

[48] The following are some examples of complex syllabic patterns that are retained throughout the poems: XX: 8, 8, 8, 8, 3, 12; XXI: 12, 10, 9, 7, 5, 12; XXVII: 5, 8, 8, 12, 8, 5; XXXVII: 8, 8, 8, 4, 12, 4, 12, 8, 8, 8; L: 11, 6, 6, 6, 11, 11. Another model is to alternate a regular stanza with an intricately patterned one, as in XXX: stanza 1: 9, 9, 9, 9, 4, 10; stanza 2: 12, 12, 12, 12. Yet another places stanzas alternately as mirror images of each other, as in XVII: stanza 1: 8, 12, 8, 8, 8, 12; stanza 2: 12, 8, 8, 8, 12, 8. Some are freer and follow no set pattern at all, e.g. VII and XXXI. In both cases this freedom is in line with the poems' themes; the description of a storm, and a series of 'petites misères', respectively.

5

The *Fleurs*: Variants and Self-Quotations

THE *FLEURS*: VARIANTS AND SELF-QUOTATIONS

PROBABLY because the *Fleurs* were never published as a volume, so that no definitive version ruled out the claims of earlier drafts, many of the poems that were intended to make up this collection survive in manuscript at various stages of composition.[1] They give a valuable insight into the way in which Laforgue wrote poetry, and we have had occasion to mention them already at various points. There are frequently three or four variants for a line, and occasionally there are as many as seven. A later version of several passages and individual lines also exists, of course, since Laforgue incorporated elements from this unpublished volume into the last poems that he wrote, the posthumously entitled *Derniers Vers*. This chapter will aim to examine both Laforgue's methods of composition and the role that these elements from the *Fleurs* play in the *Derniers Vers*.

The manuscripts of the *Fleurs* show three distinct stages of composition: an early rough draft in black ink; a copy of this on larger, smoother paper, also in black ink; and a later bold and clearly written copy in red ink on still larger sheets of more expensive paper, paper that Laforgue described proudly in a letter to Kahn as Japanese vellum.[2] Changes occur at all stages, and, in the later poems of the *Fleurs* particularly, even the last draft is frequently altered in a rough hand, either in black ink or in pencil, with considerable freedom. It is immediately clear that one-time final compositions were not treated by Laforgue with the respect that some poets might ascribe to them.

The first *brouillons* are probably the most interesting manuscripts. Here, in a tiny, spindly, and presumably fast-moving hand, we see ideas crowding in. Alterations are frequent above and below the line and in the right-hand margins, and are sometimes scrawled diagonally across a page. Humorous drawings, executed with considerable skill, often lighten the impact of a less than cheerful text. The drafts no doubt

[1] The manuscripts are in the Bibliothèque Jacques Doucet, Paris.
[2] *LA* 170.

followed each other in quick succession, since, as we have seen, the bulk of the poems must have been written within the space of about six weeks, from April to early June 1886, in a burst of poetic energy, while Laforgue lingered on in Berlin, planning his return to his literary friends in Paris, but perhaps even more preoccupied by the feelings inspired in him by his, as yet, purely formal relationship with Leah Lee.

We expect to find in the variants of almost any poet a greater tautness and economy as the writing develops, and Laforgue is no exception. He is seen practising what he had already preached to the poetess Mme Mültzer in March 1882: 'pour que cela soit plus *vision*, commencez non par: "Je rêve d'un pays . . ." mais par: "Un pays".'[3] Indeed, at times he follows his own advice with near parodic closeness: 'Ce qu'on rêve ou ce qui n'est plus' becomes 'O rêve, ou jamais plus',[4] while 'J'ai eu ce rêve dans ma sieste' becomes 'Quel lys sut ombrager ma sieste?'[5]

And perhaps we also expect to find the shift away from the abstract, and the move towards the concrete world of the senses, which is clearly visible throughout these variants. The poetic task facing the disciple of Schopenhauer and of Hartmann is rather particularly, one might think, to effect such a transition. Certainly, the particular and the tangible repeatedly replace the abstract, as the following examples will show. 'Peinte d'azur, de blanc, de rose' becomes 'Peinte d'azur, de *lait doux*, de rose';[6] 'Sa blancheur me protège contre tout ce que j'aime' is changed to 'Son *froment* me protège contre les *chairs* que j'aime';[7] and 'du moins leur expliquer' to 'Oh! *monter* leur expliquer'.[8] Particularly striking is the increased dynamism of verbs. Thus, 'je ne vois' gives way to 'je ne *tiens*';[9] 'Croyez ma bouche' becomes '*buvez* ma bouche';[10] 'Aimons-les et puis chantons-leur: Merci, c'est bien' becomes '*Cueillons en*';[11] 'Nous passons, Elle demeure' becomes 'Nous *jouissons*'.[12] Their increased astringency is also to be noted: thus, in 'Saignons tandis qu'elles exhalent | Leurs serres de Beauté, pétale par pétale', the predictable 'exhalent' is changed to the much more forceful '*déballent*';[13] and the line 'Qu'on le leur chante et leur rabâche' becomes 'Qu'on le leur *râle* et leur rabâche'.[14] The change introduces a harsher note in meaning and, interestingly, also in sound patterning. Increased phonetic patterning is not unusual. For example: 'Tu ne veux pas être fraternel'

[3] *AH* 764. [4] 'Mettons le doigt sur la plaie'. [5] 'Cythère'.
[6] 'Dimanches XXX'. [7] 'Le Brave, Brave Automne'.
[8] 'Dimanches XXXVIII'.
[9] 'Dimanches XXI': 'Je ne tiens que des mois, des journées et des heures.'
[10] 'Notre petite compagne'. [11] 'Ève sans trêve'. [12] 'Esthétique'.
[13] 'Petites Misères de juillet'. [14] 'Esthétique XLII'.

is changed to 'Tu ne veux pas *fleurir* fraternel'.[15] The play on sound in this second version emphasizes the verb's unexpectedness. Laforgue's wit is thus seen to enter the poem at a late stage.

This will surprise only those who conceive of poetry as directly received inspiration. But the variants do offer some surprises. First, the fact that colloquialisms, which Laforgue had used so abundantly in the *Complaintes*, are clearly not the natural form of self-expression. We constantly find him replacing the literary tone of his original version by a more disjointed language that has the character of speech, involving neglect or deliberate rejection of grammar, and clearly embracing the vividness of slang and idiom. Thus 'Leurs charmants yeux sont de vrais cadrans d'émail bleu' becomes 'Leurs charmants yeux *c'est* . . .';[16] 'je souffre aujourd'hui', '*ça* souffre aujourd'hui';[17]'Octobre m'a toujours noyé dans la détresse', '*fiché* dans la détresse'.[18] Examples abound. Laforgue turns to popular language for purposes of self-mockery, and, paradoxically, this is the literary style that he described to his sister Marie in the copy of the *Complaintes* that he sent her: 'Tu reconnaîtras bien . . . le cœur de ton pauvre Jules . . . il n'a pas changé, ce cœur, il est toujours aussi gros. Il est devenu un peu plus littéraire, voilà tout.'[19]

The second surprise is the absence of irony in many of the early formulations, when it frequently appears intrinsic to the poem as we know it. For example, 'Au haut de ma belle âme' was, in the first version, 'Au *fond* de ma *pauvre* âme';[20] the 'nasillardes cloches du dimanche' were 'lamentables';[21] the title 'Figurez-vous un peu' was 'Le Véritable Amour' (apparently without ironic intent), and the 'Petites Misères d'août' were 'Grandes'. Perhaps most striking are the variants of 'Dimanches XVI', the poem that Alain-Fournier praised for what he termed its perfection, and which indeed impresses the reader by its finely controlled sentiment. Originally, it contained many versions of the following lines that function in a totally different emotional register, one lacking all restraint:

> Oh! qui nous dira quel chagrin bien noir
> T'a poussée à cet acte de désespoir
> Moi je t'aurais epousée, je t'assure
> Oh! quelle que fût ta pauvre figure.

Clearly, the ironic tone was still 'littéraire' in the sense of being artificial.

[15] 'La Vie qu'elles me font mener'.
[16] Ibid.
[17] 'Petites Misères de juillet'.
[18] 'Petites Misères d'octobre'.
[19] July 1885.
[20] 'Romance'.
[21] 'Dimanches XXVIII'.

Laforgue created an ironic effect, as can be seen in many of the above examples, by deliberate contradiction of his earlier draft. Examples of such polarities can also be found, of course, if not so commonly, in the variants of other poets, such as Baudelaire or Apollinaire. And they are natural to an ironic poet, dealing necessarily in opposition and paradox, and conscious that such oppositions are frequently more apparent than real. They emerge as central to Laforgue's method of composition. Thus we find 'Amantes!—tout et toujours Femmes' becoming 'Amantes, mais jamais ma Femme';[22] and 'C'est ma destinée immortelle et première' changed to 'ma destinée incurable et dernière'.[23] And the extreme of polarization is perhaps reached when Laforgue replaces 'noble' by 'lâche' in the following lines:

> Oh! rien qu'un lâche point d'arrêt
> Dans mon destin qui se dévide! . . .[24]

'Le semblable, c'est le contraire',[25] we seem to hear Laforgue muttering with his Lord Pierrot, and, indeed, throughout the variants we notice his readiness to be led in directions that appear, on the face of it, unexpected and illogical. We see what is certainly a highly inventive—and could be regarded as a flippant—method of composition, were it not that it clearly reflects the mobility of a personality aware of its fragmentation, of a poet struggling to come to terms with that dimension of internality whose exploration has, since the Romantics, been a particular concern of French poetry. The poet no doubt does not even wish to examine analytically the oppositions and discontinuities in the processes of the mind, for 'L'artiste s'agite: l'Inconscient le mène.'[26]

For us, however, it is interesting to follow the progression even of individual lines. In 'Mettons le doigt sur la plaie' Laforgue first wrote: 'Je pleure de passé, je rage d'espérance'. 'J'exulte d'espérance' replaced 'je rage d'espérance', so that we get: 'Je pleure de passé, j'exulte d'espérance'. The parts of the whole are then reversed, in a technique that obviously also produced the 'succulent' eyes and 'sad' mouth of Elsa in *Lohengrin*: 'Exultant de passé et pleurant d'espérance'. Finally, Laforgue reached an expression that owes something to each earlier attempt, but discovers a new vocabulary: 'Tout vibrant de passé, tout pâle d'espérance'. Ironically, like so many of the best formulations, it has every appearance of spontaneity.

[22] 'Maniaque'. [23] 'Figurez-vous un peu'.
[24] 'Complainte des crépuscules célibataires'.
[25] 'Complainte de Lord Pierrot'. [26] *MP* 196.

This is to look at the changes from the semantic angle. If we consider the changes from the rhythmical and phonetic angles, we observe that Laforgue moves from a totally symmetrical line to the more telling slight asymmetry of the final version. He keeps the central caesura in all versions, but wavers between the iamb and the anapaest in stress patterns, ending on a combination of the two. When he uses the iamb, he follows it in all cases by a succession of three unstressed syllables, the first of which is always an unaccented *e*. The fact that he accepts this pattern at all shows how slight was the value he placed on this syllable even before his move to free verse. In all versions he retains different permutations of the alliteration on the letter *p*. Nasal assonance on stressed syllables appears in the last two versions, and the final version is richer phonetically than any preceding version, with the inclusion of the repeated 'tout'. No doubt this progression, in which we see sound and meaning move more closely in harmony with one another, was, like the semantic one, more instinctive than we make it appear.

Perhaps the journey towards the sensitive rendering of inner feelings can best be illustrated by taking an entire stanza (from 'Dimanches XLIV') and comparing the first and final versions:

1. J'erre, épris du fond de mon sang, d'une hostie
En fine et rougissante chair,
Qui va fièrement sertie
Dans une toilette d'hiver
Errer le long des cris sauvages de la mer. . .

2. J'aime, j'aime de tout mon siècle! cette hostie
Féminine en si vierge et destructible chair
Qu'on voit, au point du jour, altièrement sertie
Dans de cendreuses toilettes déja d'hiver,
Se fuir le long des cris surhumains de la mer!

In the final version that typically Laforguian accuracy of depiction, 'l'expression parfaite . . . de quelque chose',[27] as Alain-Fournier described it, is much more in evidence. The introduction of an aware irony in 'de tout mon siècle', the increased sexual resonance of 'vierge et destructible' (replacing the more simply descriptive 'fine et rougissante'), and the greater evocative force of each new word or phrase, whether it be 'cendreuses', 'déjà', 'se fuir', 'surhumains', or even 'si' or 'cette'—all help to create that sense of human vulnerability in an indifferent outer landscape that is one of the moods we associate most

[27] J. Rivière and Alain-Fournier, *Correspondance*, i (Paris, 1926), 143.

closely with Laforgue. We feel that it has been captured by listening to the 'imprévu' of the inner voice, and Laforgue is more than ready to grant this in a passage in which he links the poetic process with the painting of a picture or the composition of a piece of music:

On peut hardiment l'avouer: une poésie n'est pas un sentiment que l'on communique tel que conçu avant la plume. Avouons le petit bonheur de la rime, et les déviations occasionnées par les trouvailles, la symphonie imprévue vient escorter le motif; tout comme un peintre est amené là—à ce gris perle à propos de bottes, à ce géranium sans nécessité . . . tel le musicien avec ses airs qui ont l'air parasites.[28]

Already, in the spring of 1882, Laforgue wrote: 'Je deviens . . . mallarméen', and thought he had similar poetic goals: poetry that would represent 'de la psychologie dans une forme de rêve'.[29] On the evidence of variants, however, Laforgue's method of composition could scarcely offer a stronger contrast to that of the 'master'. If one takes the two well-known versions of 'Le Pitre châtié' as an example of Mallarmé's method, one sees that Mallarmé's final version is verbally no closer to the first than Laforgue's are to his. The reverse is in fact the case. But the direction of the poem has been maintained, even if the much later second version acquires, as Leo Bersani argues,[30] a different 'status' by creating an ironic distance between itself and the simpler intuitions that formed the substance of the first version. What Mallarmé has done, is to render an obsessive vision with the more accurate and complex ambiguity that it required. While Laforgue was also, in his way, a verbal perfectionist, an instinctive manipulator of language, his approach to composition appears to have been quite different, a fact that he seems to have recognized when he wrote of Mallarmé: 'Sa technique est . . . raisonnée, consciente et l'on voit souvent qu'elle n'est pas de premier jet.'[31] Nor do we have the 'premier jet' much of the time with Laforgue, one is tempted to reply to this statement. But the variants, in which we have seen the poet darting in several directions, each in a sense represent a fresh start, and many suggest other poems that might have been written. With youthful impetuosity, Laforgue moved on.

We have seen first versions gain in tautness, vividness, wit, accuracy in the portrayal of complex states of mind. But we have not, of course, been examining a progression to a state that satisfied the poet. These

[28] *MP* 129-30.

[29] *AH* 757.

[30] L. Bersani, *The Death of Stéphane Mallarmé* (Cambridge, 1982), 19-24.

[31] *MP* 129.

'final versions' were themselves dismissed. Laforgue was both prolific and self-critical, his attention constantly directed towards future projects. It is in character for him to be dismissive of his recent compositions, but the *Fleurs* constitute something of an exception. While his attention focused with intense concentration on the 'discovery' of free verse, he still had a use for these unpublished poems. Passages from many of them were incorporated into the *Derniers Vers*, where they contrast strikingly with the free verse of the volume, and where their function is, not surprisingly, frequently an ironic one. Before considering the many influences that contributed to the shaping of the *Derniers Vers*, therefore, we shall pause to consider these 'self-quotations' from the earlier volume. As we do this, we shall at times seem to enter the territory of the *Derniers Vers* themselves, but our concern will be the elucidation of the role assigned to these poems from the *Fleurs*.

Two 'uses' of the *Fleurs* are immediately clear: either a section of dramatic narrative may form the core of a new poem, or else dramatic moments, high points from various poems, first or last stanzas, may assist in the creation of a new whole. They are frequently stanzas whose regular rhythm forces itself insistently on the reader. Laforgue wrote to Kahn in a now famous letter of July 1886: 'J'oublie de rimer, j'oublie le nombre des syllables, j'oublie la disposition des strophes, mes lignes commencent à la marge comme de la prose. L'ancienne strophe régulière ne reparaît que lorsqu'elle peut être un quatrain populaire etc.'[32] And if the precise quatrains are not retained (Laforgue frees them by the addition or subtraction of syllables, often for blatant humorous effect), nevertheless the gist of Laforgue's statement is accurate. The *Fleurs* are essentially used in opposition to Laforgue's new free style, and appear deliberately archaic in their altered context.

'Solo de lune' is perhaps the most obviously successful poem in the new style that is built around a poem from the *Fleurs*. At its centre is 'Arabesques de malheur', an autobiographical description of past experience:

Nous nous aimions comme deux fous;
On s'est quittés sans en parler.
(Un spleen me tenait exilé
Et ce spleen me venait de tout.)

[32] *LA* 193.

Que ferons-nous, moi, de mon âme,
Elle de sa tendre jeunesse!
O vieillissante pécheresse,
Oh! que tu vas me rendre infâme!

Des ans vont passer là-dessus;
On durcira chacun pour soi;
Et plus d'une fois, je m'y vois,
On ragera: 'Si j'avais su!'. . .

Oh! comme on fait claquer les portes,
Dans ce Grand Hôtel d'anonymes!
Touristes, couples légitimes,
Ma Destinée est demi-morte!. . .

—Ses yeux disaient: 'Comprenez-vous!
'Comment ne comprenez-vous pas!'
Et nul n'a pu le premier pas;
On s'est séparés d'un air fou.

Si on ne tombe pas d'un même
Ensemble à genoux, c'est factice,
C'est du toc. Voilà la justice
Selon moi, voilà comment j'aime.

If the poem has limitations, it is because of a certain starkness in the presentation of a single experience, and a seemingly arbitrary condensation of material that robs this experience of a context in which it can operate. In 'Solo de lune', while scarcely altering the original poem, Laforgue adds to it the natural flow of another experience, a night journey. Two main elements are thus set in apposition: a subjective view of a past relationship, and an environment that invites reflection. The setting that Laforgue provides allows the poet's reminiscence to fall into place as the fragment of experience that it is:

La lune se lève,
O route en grand rêve!
O route sans terme,
Voici le relais,
Où l'on allume les lanternes,
Où l'on boit un verre de lait,
Et fouette postillon,
Dans le chant des grillons,
Sous les étoiles de juillet.

. . . .

Voici qu'il fait très très-frais,
Oh! si à la même heure,
Elle va de même le long des forêts,
Noyer son infortune
Dans les noces du clair de lune!. . .
(Elle aime tant errer tard!)
Elle aura oublié son foulard,
Elle va prendre mal, vu la beauté de l'heure!
Oh! soigne-toi je t'en conjure!
Oh! je ne veux plus entendre cette toux!
Ah! que ne suis-je tombé à tes genoux!
Ah! que n'as-tu défailli à mes genoux!
J'eusse été le modèle des époux!
Comme le frou-frou de ta robe est le modèle des frou-frou.

The magic of the starlit night recalls the magic of past love, and isolates the submerged feelings of regret. Under its soothing influence, bitterness is forgotten and the renunciation of 'Arabesques de malheur' has lost its appropriateness. A tenderness, only briefly glimpsed in 'Arabesques', rises to the surface, and, in place of the poet who was eager to tell of an experience and to justify his actions (or absence of them), we have one who is moved by an atmosphere into recollection and wonder.

The poem has three closely interwoven tones: tenderness and regret for the girl, calm absorption into an outside atmosphere, and an aware irony that is no longer destructive. It becomes a new and particularly acute vision of past experience, induced by an atmosphere of calm beauty. The two themes or 'phrases mélodiques',[33] the poet's love and his response to the environment, alternate throughout the poem. The strong present of the journey leads the poem, and provides its descriptive register. The love theme takes the poem inevitably into past narrative, but also into future conjecture as well as into present imaginings. Stanzas 1, 2, 3, and 5 of the *Fleurs* poem, woven into 'Solo de lune', provide the past narrative. These stanzas, quoted with minor, if highly significant, alterations,[34] appear in the first half only of 'Solo de lune'. They tell the tale, and give way, as is natural, to present imaginings in free verse.

If the large amount added to 'Arabesques de malheur' to form 'Solo de lune' is, poetically, the most remarkable part of the poem, the borrowed elements form a hard core in this free-ranging and imaginative

[33] *AH* 757. [34] See Ch. 9, section 2.

work—necessary to its final effect. The lines from the *Fleurs*, interpolations as they essentially are, distanced, since they now exist in relation to a new main text, have acquired a different function. They indicate mockingly their existence as subjective and momentary 'aperçus'. They represent one of the many voices of the narrator, one of the many ways of perceiving reality. They claim no authenticity for themselves. They are a part only of the total orchestration, and contribute to the effect of complex simultaneity in which the voices of lover, aesthete, and ironist mingle.

Derniers Vers VIII, 'Légende', also borrows heavily from a single poem of the *Fleurs*. Here the new elements frame the borrowed poem, 'Dimanches XLIV', which again provides the narrative. The *Fleurs* poem is very evocative in its brevity and disjointedness, and particularly so in the snatches of reported or imagined conversation, which are altered slightly but are essentially retained. To the narrative are added an impressionistic and dramatic new opening section, and a lyrical and, at the same time, ironic final section. Introspective meditation calls up the episode already sketched in the *Fleurs* poem, and the new poem returns again to meditation. Viewed from the angle of its development, the *Fleurs* poem has acquired perspective and depth; viewed from the angle of the finished *Derniers Vers* poem, what strikes the reader is the importance to the new whole of the 'ready-made' narrative offered by 'Dimanches XLIV', with its staccato tone and elliptical language, contrasting in register with the richer work of the later Laforgue. The following example will help to illustrate this contrast:

'Assez! assez!
'C'est toi qui as commencé!

'Tais-toi! tais-toi!
'On n'aime qu'une fois'!
(borrowed from 'Dimanches XLIV')

C'est des légendes, c'est des gammes perlées,
Qu'on m'a tout enfant enseignées,
Oh! rien, vous dis-je, des estampes,
Les bêtes de la terre et les oiseaux du ciel
Enguirlandant les majuscules d'un Missel,
Il n'y a pas là tant de quoi saigner?
('Légende')

The *Fleurs* are also used, however, in more complicated ways. *Derniers Vers* III, 'Dimanches', quotes from no less than six poems from the *Fleurs*: 'Le Vrai de la chose V', 'Dimanches XVIII', 'Célibat, célibat, tout n'est que célibat XX', 'Gare au bord de la mer XXIV', 'Dimanches XXVIII', and 'Dimanches XXX'—three poems, therefore, also entitled 'Dimanches', which are the ones first introduced. The *Derniers Vers* poem begins with elements from the last stanzas of two *Fleurs* poems (XVIII and XXX). The clinching last lines of these poems are used as an introduction to the new poem, illustrating the freedom of Laforgue's creative method. The poem continues with two new stanzas that provide a setting for the poet. These are followed by four stanzas that borrow heavily from a single *Fleurs* poem ('Dimanches XXVIII'), placing this narrative offering firmly at the centre of the *Derniers Vers* poem. Before concluding the work with a totally new section, Laforgue introduces quotations from three more *Fleurs*. (The poems used are XX, V, and XXIV—the opening stanza of XX, the first and last stanzas of V, and the last line of XXIV—again dramatic and important moments from the earlier poems.) Although Laforgue frees their metre to some extent, nevertheless these quotations contrast strongly with the final free-verse stanza. Throughout this strange amalgam we notice a pattern of alternation between the borrowed and the new, and we notice also that the sharp epigrammatic phrases taken from various *Fleurs* poems all have the analytical function of highlighting the poet's state of mind. They surround the more extensive quotation from a single poem that provides the main episode.

We have already discussed the importance of such narrative sections drawn from the *Fleurs*. They can provide Laforgue with the emotional centre of his poem. Here the section recounts the poet-narrator's changing feelings as he watches a girl return home from Mass, listens to her piano-playing in that bourgeois area from which he feels himself excluded, and indulges in sadistic fantasies of violating her innocence. All this is drawn quite simply from 'Dimanches XXVIII'. The other elements in the poem (borrowed and new)—the poet's situation, his self-analysis, his memories, his romantic dreams, his self-condemnation—arrange themselves around this episode. The question now is how the number of often very short quotations, which are gathered from various *Fleurs* poems, work. We are presented with a series of epigrammatic formulations, strung together like so many illuminating flashes, providing reminiscences and summings-up, aspirations and accusations:

> Bref, j'allais me donner d'un 'Je vous aime'
> Quand je m'avisai non sans peine

Que d'abord je ne me possédais pas bien moi-même.
(based on XVIII, final stanza)

(Mon Moi, c'est Galathée aveuglant Pygmalion!
Impossible de modifier cette situation.)
(based on XXX, final lines)

Ainsi donc, pauvre, pâle et piètre individu
Qui ni croit à son Moi qu'à ses moments perdus,
(based on XXX, second stanza)

Non, non! C'est sucer la chair d'un cœur élu,
Adorer d'incurable organes
S'entrevoir avant que les tissus se fanent
En monomanes, en reclus!
(based on XX, first stanza)

Et ce n'est pas sa chair qui me serait tout,
Et je ne serais pas qu'un grand cœur pour elle,
Mais quoi s'en aller fair les fous
Dans des histoires fraternelles!
(based on V, first stanza)

L'âme et la chair, la chair et l'âme,
C'est l'Esprit édénique et fier
D'être un peu l'Homme avec la Femme.
(based on V, final stanza)

Oh! file ton rouet et prie et reste honnête.
(based on XXIV, final line)

The likelihood, for a poet of this time, is that such disparate elements would *not* work. Even Rimbaud does not appear to have composed in quite this manner. They can be seen as inaugurating the fashionable collage technique, used perhaps most notably later by Apollinaire (who owed more to Laforgue than he cared to admit), to convey, first, a sense of discontinuity and, secondly, that of a complex simultaneity—a sense of the coexistence of different separate levels of the mind. When the final controlled and rational (new) stanza appears in the poem, and the poet is accused of madness, we leave the disordered world that we have inhabited with relief, for it has not merely been described to us, it has quite simply been flung at us, rather in the manner of the Absurd theatre of Beckett or Ionesco. Of course, there is subtle organization and a concealed order, but the impact of the poem is largely due to the fact that the discontinuities of the text directly echo or 'translate' those of the mind.

If we were to consider *Derniers Vers* IV, V, IX, and X, which also rely heavily on quotations from the *Fleurs*, we would discover broadly similar effects, increasing our sense of the intricate simultaneity of feelings that can be conveyed by this method of composition, and, consequently, of the value that these elements from a rejected volume have in their new setting. Often sharply ironic in the first place, they become doubly so, providing, as they do, an opposition between time-scales and stylistic registers, between lines originally composed in syllabic metre and lines in free-verse phrasing. This aspect can only be fully appreciated when we see them in those settings. By isolating them, two main types have emerged clearly: one is the fragile and compelling narrative, around which the complex poem of interior monologue is frequently structured, and the other is the series of fragmentary and sharply conflicting impressions and formulations, which themselves, as we have seen in the last poem, powerfully re-create the contradictions central to that monologue.

6

Towards the *Derniers Vers*: 'Trouver une langue'

WALT WHITMAN AND FRENCH CONTEMPORARIES

THE *Derniers Vers* was the first volume in French to be composed entirely in free verse, and it owes some of its fame to this fact. The poems were not published as a volume until after Laforgue's death, but eleven of the twelve appeared during his lifetime, ten in pairs in Kahn's *La Vogue*, and one in Téodor de Wyzewa's *Revue indépendante*. We have noted Laforgue's constant experimentation with metre, and his use of a *vers libéré* that was influenced by some of his contemporaries—most notably Verlaine. In free verse he was a leader, but he was nevertheless influenced by a number of factors that determined the particular direction he followed. Two stand out clearly: one was the atmosphere of avant-garde metrical experimentation, in which he participated intellectually, if not actually; and the other was the example of Walt Whitman's free verse, which he translated into a parallel French form.

Prose poetry had, since Baudelaire, offered the main alternative to metrical verse, and Laforgue had himself included one prose poem[1] in a volume of verse and had for some time used an impressionistic and poetic prose in his many descriptive passages, some of them intended perhaps for the projected novel *Saison*, which has never been found and which may never have developed beyond such fragments.[2] He referred enthusiastically to Mallarmé's prose poem 'Plainte d'automne' in his correspondence as early as January 1881,[3] but it was not until just before his move into free verse that he discovered Rimbaud's prose poetry, when Kahn was publishing the *Illuminations*, a few at a time, in *La Vogue*. The impact they made on him was immediate and powerful, as his comments show: 'Ce Rimbaud fut bien un *cas*. C'est un des rares qui m'étonnent. Comme il est entier! presque sans rhétorique et sans attaches.'[4]

1 'Grande Complainte de la ville de Paris'.

2 Published in *MP* and *FV*.

3 *AH* 693: 'Je me rappelle certain poème en prose de Mallarmé sur l'automne, voilà!' 'Frisson d'hiver', 'Un spectacle interrompu', 'Le Phénomène futur' were in the same issue of the *République des lettres* of 20 Dec. 1875.

4 *LA* 187.

One imagines that what surprised the introspective Laforgue was the power of this new poetic voice. Like Whitman, Rimbaud must have suggested to him the possibility of synthesis in poetry ('comme il est entier!'): of wholeness as against fragmentation. Rimbaud *was* the voice of that unconscious which Hartmann proclaimed and Laforgue worshipped, but whose spirit he had never properly conveyed in his self-questioning verse. And he was the voice that was new ('sans attaches', 'fleur hâtive et absolue sans avant ni après'[5]), that broke entirely with French traditional verse and with its world of accepted literary reference. Rimbaud appeared to have no antecedents, and no doubt Laforgue, whose early verse echoed his many enthusiasms, who had more recently resented the suggestion that he resembled Corbière, and whose poetry was unquestionably intertextual, admired this impressive isolation. He can scarcely have read the *Illuminations*, which he may even have seen in their entirety on a brief visit to Paris and to Kahn in June 1886, without being aware of the new kind of prose poetry that they represented, so strikingly different from that of his predecessors, including Baudelaire and Mallarmé. Laforgue, never for a moment doubting Rimbaud's poetic integrity, was convinced by the dynamism of the writing, and would probably have found a modern reading such as Todorov's,[6] which emphasizes the deliberate obscurity of the poems, difficult to comprehend. Dislocation and 'néant' were Laforgue's starting-points as much as they were Rimbaud's, if in very different ways; they were necessary evils, and it was their partial transcendence that was surprising and exciting. 'Tout est dans la richesse inouïe du pouvoir de confession, et l'inépuisable imprévu des images *toujours adéquates*',[7] he noted of Rimbaud in his jottings on some contemporary poets, characterizing the writing as 'le genre somnambule', and commenting on Rimbaud's refusal to be the slave of form, 'jamais de strophes, de factures, de rimes'.[8] Rimbaud's prose poetry must have proved to Laforgue finally and completely that the essence of poetry did not reside in verse form, for Rimbaud had written prose that quite clearly set itself all the objectives traditionally reserved for verse.

The *Illuminations* contain two poems that no one hesitates to label free verse—'Marine' and 'Mouvement'—and Laforgue would have seen these in the May and June 1886 numbers of *La Vogue*. There is no

[5] *MP* 129.

[6] T. Todorov, 'A Complication of Text: *Les Illuminations*', in T. Todorov (ed.), *French Literary Theory Today* (Cambridge, 1982).

[7] *MP* 129 (my emphasis).

[8] Ibid.

record of his reaction to them—nor, indeed, of any awareness on Rimbaud's part of their accidental novelty. Grojnowski claims that other poems in the volume—'Enfance III' and 'Départ'—are also written in free verse, and, using his criteria, 'Veillées I' could also be suggested: 'On cite régulièrement "Marine" et "Mouvement" . . . comme premiers exemples de poèmes écrits en vers libres. En fait cette forme apparaît déjà dans le numéro du 13 mai [of *La Vogue*] avec "Enfance III" ou "Départ".'[9] 'Enfance III', which is, of course, only part of a poem, the rest of which is clearly in poetic prose, contains two exceptionally long lines, both of thirty-two syllables, and this may be why it is not usually held up as free verse. Even 'Mouvement' contains two lines of more than usual length of course—one of twenty, and one of twenty-one syllables. The free-verse line stresses the 'rythme personnel'[10] of the poet—this is its main function and justification—and the shorter line is usually thought to be better fitted to this task. One could argue against 'Départ' on the grounds that three of its four lines, broken as they are by full stops,[11] do not possess the unity required of the free-verse line, which was expected at this period to be a stronger sense-unit than the line of metrical verse, since this was what justified its isolation. The total break within the line, marked by a full stop,[12] seems to deny it its role. However, as free verse developed, it made use of whatever freedoms suited the particular requirements of a poet at any given moment, and so, on occasion, included in its repertoire the break marked by a full stop within a line as well as the exceptionally long line. In 1894, some years after the 'mêlée', Mallarmé described free verse admirably as a 'modulation . . . individuelle' that allowed the poet to express his 'chant profond', since 'toute âme est un nœud rythmique'.[13]

If differences of definition persist today, it is hardly surprising that the picture was confused at the time that the changes were actually taking place. What is clear is that the move to free verse happened

9 D. Grojnowski, 'Poétique du vers libre: *Derniers Vers* de Jules Laforgue', *Revue d'histoire littéraire de la France*, 1984: 396.

10 G. Kahn, 'Préface sur le vers libre', in his *Premiers Poèmes* (Paris, 1897): 'Depuis longtemps je cherchais à trouver en moi un rythme personnel suffisant pour interpréter mes lyrismes avec l'allure et l'accent que je leur jugeais indispensables.'

11 'Assez vu. La vision s'est recontrée à tous les airs. | Assez eu. Rumeurs des villes, le soir, et au soleil, et toujours. | Assez connu. Les arrêts de la vie. Ô Rumeurs et Visions!'

12 Question and exclamation marks occur frequently, and do not represent a similar break.

13 Stéphane Mallarmé, *Œuvres complètes* (Paris, 1945), 644: 'Une heureuse trouvaille avec quoi paraît à peu près close la recherche d'hier, aura été le *vers libre*, modulation (dis-je souvent) individuelle parce que toute âme est un nœud rythmique.'

gradually and without the self-awareness that we might be tempted to attribute to it, and that certain examples of what was taken to be prose poetry, but was in fact a kind of free verse, appeared without arousing any interest in the novelty of their form. As early as 1882 Marie Krysinska published poems that could be debarred from being classified as free verse only if one considers the use of the long line a disqualification. In 'Chanson d'automne', the poem that Grojnowski says 'mérite d'être largement citée en raison de sa ligne prosodique qui, bien avant les essais des novateurs, atteste une instauration, passée inaperçue',[14] lines run to 24, 26, and 31 syllables, but the poem is in other respects in acceptable free verse. Although he knew Marie Krysinska's poetry, Laforgue does not appear to have been struck by its originality of form, concerning himself instead, not unjustifiably, with the inauthenticity of its author's rhetorical tone.[15] The use of the long line, whose antecedent is clearly the biblical verset, in both the Rimbaud and the Marie Krysinska poems is of considerable interest, since this is also the practice of Walt Whitman, who frequently uses lines running to more than twenty syllables (to transpose a French syllabic count on to verse written in English, merely in order to compare approximate lengths). Whitman's innovation, of course, found little favour with some English poets. Thomas Hardy, some twenty years Whitman's junior, wrote of the 'woeful fogs of free-verse worship', and said of Whitman that he 'wrote as he did, formlessly, because he could do no better'.[16]

Of Laforgue's contemporaries, Kahn is the poet who claimed both to have influenced him and, indeed, to have beaten him to the winning-post in the free-verse race. That he supplied the literary atmosphere that encouraged Laforgue, and offered him the opportunity to publish his first free-verse poems in *La Vogue* is not in doubt. That he was ahead of his friend in the move towards free verse is perhaps difficult to maintain in the face of Laforgue's surprise, expressed in a letter to him, at metrical innovations that did not go as far as free verse: 'Il y a là des tas de rythmes inédits *et je ne te connaissais pas ces préoccupations*.'[17] And when, 'frappé d'amnésie',[18] Laforgue 'discovered' free verse, he com-

[14] Grojnowski, 'Poétique du vers libre', 395–6.

[15] *AH* 832: 'Marie Krysinska a sensibilité artiste à fond original, mais tout cela est bien noyé de la rhéto[rique] à la mode, n'est-ce pas?'

[16] T. Hardy, *Collected Letters*, ed. R. L. Purdy and M. Millgate (Oxford, 1985), vi. 165; v. 192.

[17] *LA* 182 (my emphasis).

[18] Grojnowski, 'Poétique du vers libre', 392: 'Son désir d'échapper aux conventions du langage le frappe d'amnésie.'

municated this to Kahn with a note of urgency and exaltation, which makes little sense had he been writing to someone who was known to be doing the same: 'J'oublie de rimer, j'oublie le nombre des syllabes, j'oublie la distribution des strophes . . .'[19]

The translation of Walt Whitman's free-verse poems may well be what actually brought this amnesia about. Not only did Laforgue have the examples of powerful poems written in free verse by a major poet, but he was translating them into a parallel free-verse form. A common vehicle for translations from English into French was prose—for example, Mallarmé's translations of Poe[20]—the assumption being that French verse imposed insuperable difficulties on the translator. Some poets, Elizabeth Browning and Rossetti among them, were rendered line by line, and so, *faute de mieux*, in a kind of free verse, but no significance was attached to the form. Whitman's own free verse, however, removed the difficulty of choice, since it was obviously the natural medium, and this presumably meant that Laforgue could respond to the challenge of offering an equivalent on the stylistic as well as on the semantic level. He was having English lessons with Leah Lee at the time, and, introduced to Whitman's poems by the American-born poet Vielé-Griffin, probably translated them to help fill out the early issues of *La Vogue*. He seems to have been able to select as he wished, since he explained in a letter the reasons for choosing 'A Woman Waits for Me', a poem that he clearly admired.

Whitman, at the height of his fame, and the unknown Laforgue make a strange pair. However, parallels between the two poets can be found beneath the obvious contrasts. Both write strongly in the first person, and the spoken word is vitally important in the work of each. Both are dramatic poets, in that they juxtapose attitudes and are open to the irrational and often conflicting associations of the unconscious, which

[19] *LA* 193. Kahn is, of course, intimately connected with the development of free verse. In the spring of 1882, when Laforgue wrote: 'Je deviens (comme forme) kahnesque et mallarméen' (*AH* 757), he was clearly in advance of Laforgue in his poetic ideas. He published the first series of free-verse poems, entitled 'Intermède'—the title was to be given to a section of *Les Palais nomades*—in *La Vogue* on 28 June 1886, thus preceding the publication of any of the *Derniers Vers*. The first *Derniers Vers* to be published were 'L'Hiver qui vient' and 'La Légende (later 'Le Mystère . . .) des trois cors' (*La Vogue*, 16 Aug.). He also produced the first volume to contain a significant number of free-verse poems: *Les Palais nomades* (1887). Laforgue's *Derniers Vers* were not published as a volume until 1890. He expounded his views on free verse in his famous preface to his own poems, 'Préface sur le vers libre' (*Premiers Poèmes* (Paris, 1897)), and as early as 1888 he published an article on free verse in answer to Brunetière (see P. Mansell Jones, *The Background of Modern French Poetry* (Cambridge, 1951), 126).

[20] Shelley, Keats, and Swinburne were similarly translated.

they both revere. Nevertheless, Whitman's hymn to sexuality, procreation, and the future in the *Leaves of Grass* could scarcely contrast more strongly with the self-questioning ironic verse of Laforgue: on the one hand, poems that proclaim the ecstasies of the flesh—and the spirit—on the other, poems of youthful frustration and yearning. 'A presence that is a sexual self-knowledge'[21] is Harold Bloom's accurate description of Whitman, while Whitman provides his own:

> Walt Whitman, a Kosmos, of Manhattan the son,
> Turbulent, fleshy, sensual, eating, drinking and breeding.[22]

They were not so different, however, that Laforgue could not appreciate the American poet. Sending his translation of 'A Woman Waits for Me' to Kahn, he wrote: 'C'est un des plus beaux Whitman du volume'; and this was said despite the fact that Whitman's ideal woman has a New World forthrightness, athleticism, and lack of modesty that have little in common with Laforgue's delicate evocation of that hothouse plant, the bourgeois, sexually unawakened 'jeune fille'. He added: 'Je crois l'avoir très heureusement traduit'[23]—'heureusement', not 'fidèlement', suggesting the creativity that he saw as a necessary part of his task. And in fact the translation captures very precisely the full-blooded original. Pains have clearly been taken not to omit or distort elements, and happy near-equivalents have been found for 'untranslatable' phrases. Laforgue nowhere alters the distribution of lines, keeping the Whitmanesque long line, and keeping also those balanced repetitions which are a common feature of the free verse of both poets. When the sense seems to make this desirable, he expands on a bare translation of the English, quite legitimately adding words to match the rhetorical force of the original.[24] Where phrases need transposing into a different idiom, he tackles the problem squarely.[25] There are occasional misunderstandings—and therefore mistranslations[26]—and one apparently deliberate

21 H. Bloom, *Agon* (New York, 1982), 187.

22 W. Whitman, 'Song of Myself', *Leaves of Grass*.

23 *LA* 197.

24 '. . . and not only for our own sake but for others' sakes' becomes: 'non seulement pour l'amour de nous, mais pour l'amour d'autres *encore*'.

25 'Women that are warm-blooded and sufficient for me' is translated as: 'Ces femmes qui ont le sang chaud et *peuvent me faire face*.' 'I am stern, acrid, large, undissuadable, but I love you' is translated as: 'Je suis austère, âpre, immense, *inébranlable*, mais je t'aime.' 'I press with slow rude muscle' is translated as: '*J'y vais* d'un muscle rude et *attentionnée*.'

26 'I shall look for loving crops from the birth, life, death' is translated as: 'Et je *surveillerai* les moissons d'amour, naissance, vie, mort.' 'I brace myself effectually' is translated as: 'Et je *m'enlace* bien efficacement.'

change from the unanimistic plural 'all women' to the singular in the 'tu' form:

I draw you close to me, you women,
I cannot let you go,

Je t'attire à moi, femme,
Je ne puis te laisser passer,

which significantly and interestingly narrows the context of the poem. But if one also includes the translations of other Whitman poems,[27] one must conclude that Laforgue is a careful, sensitive, and inventive translator. Occasionally he fails to reproduce the lyrical power of the English, particularly when this depends on complex sound patterning. Thus, it is scarcely Laforgue's fault that:

A travers toi, je lâche les fleuves endigués de mon être,
En toi je dépose un millier d'ans en avant,

does not fully render Whitman's splendid lines:

Through you I drain the pent-up rivers of myself,
In you I wrap a thousand onward years.

If Laforgue was never to write with a similar boldness, never to write 'of life immense in passion, pulse and power',[28] the reading and translating of such 'strong' poetry cannot have left him unaffected. 'L'Hiver qui vient', the first free-verse poem that Laforgue wrote, contains, as well as a hint of unanimism, what is probably his most positive ending:

Tous les ans, tous les ans,
J'essaierai en chœur d'en donner la note.

It is still, of course, a long way from 'Whitman's unremitting wish to end poems in the affirmative'.[29] The poem also contains a higher proportion of long lines than any other of the *Derniers Vers*. Perhaps even more revealing of a Whitman influence is the rhythmical, almost liturgical, use of anaphora, of a kind that is rarely found in the *Fleurs*, but is at the heart of Laforgue's, as of Whitman's, free verse.[30] Some lines

27 These were: 'One's-Self I Sing', 'To Foreign Lands', 'To a Historian', 'To a Certain Cantatrice', 'Shut Not your Doors', 'Poets to Come', 'To You', 'Thou Reader' (all Inscriptions), and 'O Star of France'.

28 W. Whitman, 'One's-Self I Sing'.

29 H. Vendler, *The Music of What Happens* (London, 1988), 141.

30 This anaphora must not be confused with the use of the refrain in his metrical verse.

from 'A Woman Waits for Me' and 'L'Hiver qui vient' will help to illustrate this point. The repetition is sometimes no more than a pronoun, followed by a parallel sentence structure:

> I draw you close to me . . .
> I cannot let you go . . .
>
> Et il gît là,. . .
> Et il frissonne. . .

Even such slight repetition illustrates the linearity of this early free-verse composition. Sometimes repetition is more extended:

> I see that they understand me and do not deny me,
> I see that they are worthy of me, I will be the robust husband of
> those women.
>
> C'est la toux dans les dortoirs du lycée qui rentre,
> C'est la tisane sans le foyer . . .

The effect is to add a sense of accumulation to the linear composition. Repetition is most often found, as here, at the start of lines; the lines are end-stopped, and this also is a common feature of Whitman's, as of Laforgue's, early free verse.

In addition to this liminal repetition, a more complex interweaving of repeated words is typical of the free verse of both poets, and can perhaps best be observed here in the final few lines of each poem.

> I shall demand perfect men and women out of my love-spendings,
> I shall expect them to interpenetrate with others, as I and you
> interpenetrate now,
> I shall count on the fruits of the gushing showers of them, as
> I count on the fruits of the gushing showers I give now,
> I shall look for loving crops from the birth, life, death,
> immortality, I plant so lovingly now.
>
> Non, non! C'est la saison et la planète falote!
> Que l'autan, que l'autan
> Effiloche les savates que le Temps se tricote!
> C'est la saison, oh déchirements! c'est la saison!
> Tous les ans, tous les ans,
> J'essaierai en chœur d'en donner la note.

Whitman's repetitions are more regular and have a more clearly oratorical function; Laforgue's are arguably subtler and, of course, contain irony. Both poets would seem to be responding to a need for increased anaphora to conclude the poem, rather as earlier poets might have turned to rich rhyming.

The rhythmical texture of Laforgue's poem, and the question of the 'Wagnerism' which was fashionable at the time will be considered later. We are concerned here simply with a similarity in the structuring of the verse: the lines emerge strongly as units, they follow one another like a series of waves, creating a sense of accumulation, and they employ anaphora as one of the means by which the 'flexion perpétuelle'[31] of free verse acquires the musicality essential to all poetry. Of course, such use of anaphora is in no way unique to Whitman and Laforgue. Rimbaud and Marie Krysinska—to name but two of the poets already mentioned—sometimes employ a similar technique. But, since translation brings a writer into very close contact with the actual structure of the verse, it seems likely that, at a moment when other forces contributed to Laforgue's receptivity, he may have been most directly influenced by his intimate wrestling with Whitman's free-verse compositional methods.

'L'Hiver qui vient' contains no elements from the *Fleurs*, and is probably the *Derniers Vers* poem whose style is most strikingly different (despite similarity of theme) from that of any other work in that volume. It represents, as Laforgue felt it did, a sudden reorientation; one that was inseparable from its metrical liberation. But Laforgue perhaps owed more to Whitman than he realized, because freedom of metre and a certain boldness and expansiveness in the treatment of subject-matter seem to go hand in hand. In this poem, longer than any of the *Fleurs*, Laforgue creates his mood more powerfully than in his earlier verse. It is not a mood with which Whitman would have had much sympathy. He writes of the approach of winter, the season of death, while Whitman's poetry is a celebration of life and of the future, carrying the reader to a mood of undeterred confidence. But Laforgue's rhetoric, despite its total dissimilarity of intent, is equally convincing. It is as if Whitman had helped Laforgue to overcome not the 'style rhéto', which he had long since abandoned, but the more recent 'style littéraire', the convoluted self-doubt that mocked his suffering almost before this had found expression. This is a poem that speaks of suffering with knowledge and clarity, and without either over- or understatement. It is at the same time a witty and richly intertextual poem, and it follows a complex and varied stream of consciousness, but when it speaks of 'toute la misère des grands centres' it is undoubtedly with an authority that had

[31] Kahn, 'Préface sur le vers libre'.

been lacking in the early work. It tells us something about one aspect of the human condition, as Whitman's poem does about its reverse aspect. Laforgue writes as a decadent and depicts the season of death, while Whitman is advocating a return to nature and spontaneity.

The cult of naïvety in Whitman—as in Rimbaud—was, of course, a conscious escape from civilization—from its materialism and its decadence. The concept of prelapsarian innocence can exist only after the Fall. Nevertheless, poets who did not appear to question the validity of either their emotions or their methods, who were unashamedly 'serious', allowed Laforgue to become more expansive and to develop the lyrical strain in his own writing. They helped him towards truer self-expression. For, despite a deliberate pursuit of novelty, the creative figures in Laforgue's immediate circle, Kahn and Bourget, lacked any real poetic originality, oppressed as they were by what might be termed, with Harold Bloom, the 'anxiety of their belatedness'.

Laforgue remained loyal to his admiration for Bourget, although the latter considered him by now to have taken many a false turn. Indeed, he was to illustrate his total incomprehension of Laforgue's aims in a letter to G. Jean-Aubry after Laforgue's death: 'J'ai beaucoup aimé Jules Laforgue et je serais très disposé à écrire quelques pages sur lui. Ce qui m'a toujours empêché jusqu'ici, c'est mon déconcertement devant la forme si peu traditionnelle qu'il avait adoptée'[32]—a truly remarkable statement for a supposed innovator. In this context, the liberating genius of Whitman, as of Rimbaud—of poets who were prepared to be iconoclastic in order to be innovative—arguably made as important an impact on Laforgue as he was later to make on the young Eliot. In both cases the experience represented an 'écart'. The liberation came with the force of revelation, but what was revealed, though previously unsuspected, was immediately and instinctively recognized as the necessary way forward. Laforgue by this time needed to become less insistently and cerebrally ironic; Eliot to have the subtle intricacies of the ironic mode 'modernized' for him so that they could serve as a vehicle *for* the expression of feeling.

In 1905 Vielé-Griffin argued that Whitman's influence on the development of French free verse had been negligible. Conveniently using his pen-name of François de Poncher, he asked what could be more different from the fluid yet analytical free verse of a Vielé-Griffin than 'l'énorme phrase insoucieuse et incorrecte de Walt Whitman, qui se prolonge en trois et quatre lignes de texte, veule et brutale . . . sans

[32] G. Jean-Aubry, MSS.

assonance, souvent sans rythme appréciable, toujours presque, sans chanson'.[33] Patently unjust to the originality and craftsmanship of Whitman's work, this is also somewhat strange in view of the fact that Vielé-Griffin singled out Laforgue, the author most closely linked to Whitman, as the writer responsible for introducing the French to free verse. 'L'œuvre de Laforgue' he wrote, 'est la clef de voûte du pont jeté entre le passé et l'avenir.'[34] He also quite correctly placed Verlaine, 'créateur de nouveautés exquises',[35] directly in line behind Laforgue: 'Laforgue peut venir, la besogne est prête.'[36] And he renamed free verse appropriately 'le rythme *nécessaire*' (my emphasis), insisting on 'la naissance simultanée de l'idée et de son expression', and thus providing us with the delightful image of a 'muse gémellipare'.[37]

IMPRESSIONISM AND THE *MORALITÉS LÉGENDAIRES*

A poet does not react merely to the work of other poets. His style may develop and change as a result of his interest in other art-forms or of his writing in other genres. Both these factors were present in Laforgue's case. Throughout the time when he was moving towards his final poetic style he was also composing short stories, to which he attached considerable importance. And if a sophisticated interest in the visual arts, particularly painting, had been a constant factor since his Parisian days, it is perhaps most significant at this time, when a clear parallel can be established between the methods and aims of the art he particularly admired—that of the Impressionists—and his own liberated compositional methods.

A brief survey of the many strands that contributed to Laforgue's expertise in the visual arts may be useful here. He had had private drawing-lessons as a child and several witty sketches survive as testimony to his skill. (Many are reproduced in the latest edition of his works, referred to here as *AH*.) After failing his *baccalauréat* three times, he turned to the École des beaux-arts, where, like his older brother Émile, he studied under Henri Lehmann and met the painter Seurat. His first job in Paris was that of secretary to the art-collector Charles Ephrussi, one of Proust's models for Charles Swann, who was preparing his edition of Dürer's drawings, and it was in his study, lined with paintings by Manet, Monet, Renoir, Degas, Pissarro, Sisley, and

[33] F. Vielé-Griffin, in *L'Occident*, 40 (Mar. 1905), 118. [34] Ibid. 120.
[35] Ibid. [36] Ibid. [37] Ibid. 122.

Berthe Morisot,[38] among others, that he became familiar with Impressionist painting and made the acquaintance of Manet. While a student in Paris, he attended Taine's lectures on art, which were later summarized to form his *Philosophie de l'art*, took copious notes, some of which survive in manuscript, and arrived ultimately at a position strongly opposed to Taine's, which he put forward in his notes on Impressionism and, less directly, in 'L'Art moderne en Allemagne', and in his contributions to the *Gazette des beaux-arts* and its supplement, the *Chronique des arts et de la curiosité*. At moments of creative discouragement he considered the possibility of making his name as an art critic.

The suggestion that a connection can be traced between Laforgue's interest in painting in general, and in Impressionism in particular, and his poetic style and ambitions is not an idea that would have seemed strange to his contemporaries. Baudelaire's views on the relatedness of the art-forms—the visual arts, literature, and music—were by now common currency, and his famous essay *Le Peintre de la vie moderne* was interpreted with poetry as well as the visual arts in mind. Zola freely acknowledged a literary debt to Impressionist painting: 'Je n'ai pas seulement soutenu les Impressionnistes', he wrote, 'je les ai traduits en littérature, par les touches, notes, colorations, par la palette de beaucoup de mes descriptions,' and, significantly stressing the fact that Impressionist painting contributed to literary innovation, he confessed: 'Dans tous mes livres . . . j'ai été en contact et échange avec les peintres.'[39]

Laforgue was convinced of the naturalness of applying the methods of the visual arts to literature. In one passage, the importance of which cannot be over-stressed, he contrasted the Impressionists' 'œil naturel'[40] with traditional academic vision, dominated by concepts that he considered artificial, such as line, relief, and perspective. The natural —or Impressionist—eye saw the external world as it really appears, that is, in multiple gradations of prismatic colour:

> où l'académique ne voit que la lumière blanche, à l'état épandu, l'impressionniste la voit baignant tout non de morte blancheur, mais de mille combats vibrants, de riches décompositions prismatiques. Où l'académique ne voit que le dessin extérieur enfermant le modelé, il voit les réelles lignes vivantes sans forme géométrique mais bâties de mille touches irrégulières qui, de loin, établissent la vie. . . . L'impressionniste voit et rend la nature telle qu'elle est, c'est-à-dire uniquement en vibrations colorées.[41]

[38] *AH* 718. [39] E. Hertz, *Europe*, 83-4 (1952), 32. [40] *MP* 133.

[41] Ibid. 136-7. For an account of Laforgue's aesthetic theories and debts, notably to Helmholtz, see M. Hannoosh, 'The Poet as Art Critic: Laforgue's Aesthetic Theory', *Modern Language Review*, July 1984: 553-69.

The painting that was Laforgue's favourite in Ephrussi's collection, Monet's 'Baignade à la Grenouillère' (now in the National Gallery),[42] certainly offers a picture of nature—and particularly of the reflective surface of water—'uniquement en vibrations colorées' by means of 'mille touches irrégulières'. These were not achieved with ease, any more than was Laforgue's apparently spontaneous verse. Monet referred to the painting as a 'pochade', though it is in fact somewhere between a rough draft and a finished painting. The use of X-rays has given us a different and much more painstaking image of Impressionist techniques. It has shown how even this painting was much altered during composition, and that it contains surprisingly elaborate combinations of coloured pigments. The Impressionists understood that the human experience of colour is not of single, determinate hues, but of a complex shifting process of interaction. Awareness of this fact led to the emphasis they placed on relationships rather than on objects, a perception that was also central to Laforgue's view of his artistic material. Interestingly, Seurat referred to the technique that he invented not primarily as 'pointillisme', but as 'divisionisme', a term that also highlights relationships rather than objects.

Laforgue drew a parallel between the Impressionist sensitivity to the variety and complexity of visual experience and the 'symphonic' ambitions of Wagner, linking both to the Hartmannian unconscious law governing the universe: 'Plus de mélodie isolée, le tout est une symphonie qui est la vie vivante et variante, comme "les voix de la forêt" des théories de Wagner en concurrence vitale pour la grande voix de la forêt, comme l'Inconscient, loi du monde, est la grande voix mélodique, résultante de la symphonie des consciences de races et d'individus.'[43] Finally, he stated that the central principle behind the Impressionist school had been adopted by writers: 'Ce principe a été, non systématiquement, mais par génie appliqué en poésie et dans le roman chez nous.'[44] No doubt he had Zola and Huysmans in mind for the novel.

As in the case of Baudelaire's essay *Le Peintre de la vie moderne*, much that Laforgue says about the visual arts in his notes on Impressionism can be applied to literature, and it may well be that his study of painting helped him to develop and refine his general aesthetic principles. As Patrick Heron has said, 'painting is the greatest externaliser of feeling; it renders perceptions actual.' Laforgue's admiration for the Impressionists' fidelity to their perceptions, and for their translation of them into

[42] 'Mes amitiés au Claude Monet que vous savez' (*AH* 712). Henri Peyre concludes incorrectly from this statement that Laforgue knew Monet (H. Peyre, *Qu'est-ce que le symbolisme?* (Paris, 1974), 176).

[43] *MP* 137–8.

[44] *MP* 138.

the medium of paint ('something stated out there in the daylight'[45]), no doubt encouraged his similar translation of perceptions into language. When he writes in his art criticism: 'l'objet et le sujet sont donc irrémédiablement mouvants, insaisissables et insaisissants',[46] one feels it is the poet as much as the art critic speaking, and the same could be said of the following highly significant fragment, also from his art criticism: 'Chaque homme est selon son moment dans le temps, son milieu de race et de condition sociale, son moment d'évolution individuelle, un certain clavier sur lequel le monde extérieur joue d'une certaine façon. Mon clavier est perpétuellement changeant et il n'y en a pas un autre identique au mien. Tous les claviers sont légitimes.'[47] Moral criteria are therefore clearly rejected in the evaluation of art. ('La morale n'a rien à voir avec l'art pur, pas plus qu'avec l'amour pur.'[48]) The mistake of Taine and his followers was to seek 'par des voies morales, littéraires, spiritualistes, l'idéal plastique'.[49] Like Baudelaire, Laforgue insisted on modernity: 'Il s'agit de n'être pas médiocre. Il faut être un nouveau.'[50] Only in this way could art be interesting, which was of primary importance. He rejected mimetic art; not only was it impossible ('l'œuvre ne sera jamais l'équivalent de la réalité fugitive'[51]), but also it denied the relative nature of 'la vie incessament ondulatoire'[52] and of the human being, that 'créature incomplète et éphémère'.[53] So Laforgue appreciated Impressionism, seeing in it, as Proust was later to do, the 'varied landscape of the hours'. 'There is no better writing on Impressionism than Laforgue's: his excess seems bound up with his powers of description,' writes T. J. Clark,[54] responding to Laforgue's insight, to his receptivity to the new, as well as to his enthusiasm. Of course, what Laforgue brought to his appreciation of Impressionism—taste and sensitivity apart—was the understanding provided by a parallel endeavour, while what he learnt most importantly from Impressionism was to trust the complex and shifting impressions of life that constituted his subjective 'perceptions', and to consider them as natural and valid—indeed, the only legitimate—material for his art.

The 'modern', the ephemeral, the fragmentary, the inconsequential were the realms of the Impressionist painter, as they were also of Laforgue the poet. Impressionist paintings offered glimpses into human

45 P. Heron, 'Late Picasso', *Modern Painters*, 1 (Summer 1988), 6.

46 *MP* 141. 47 Ibid. 48 Ibid. 152. 49 Ibid. 154.

50 Ibid. 153. 51 Ibid. 140. 52 Ibid. 139. 53 Ibid. 152.

54 T. J. Clark, *The Painting of Modern Life: Paris in the Art of Manet and his Followers* (London, 1985), 17.

existences. Manet, for example, was accused of painting pictures that simply could not be decoded. Had the duel in *Le Déjeuner* taken place or not? What was the relationship of the two women in *Le Balcon*? Was the waitress in *Un bar aux Folies Bergère* a prostitute or a respectable employee—or something in between? Integration was refused both at the narrative and at the pictorial level. A contingent world was presented that was indecipherable, while it was simultaneously recognizable and highly suggestive. The topics selected, echoing the rootlessness and anonymity of life in the modern city, appeared arbitrary, having no known before or after. They were, in Laforgue's words, 'sans nécessité'.[55]

In a significant passage Laforgue explained why these, rather than classical subjects, attracted the modern sensibility:

> Littérairement, avec des goûts d'historien, d'antiquaire, nous pouvons être amoureux sincèrement d'un type de femme du passé, Diane chasseresse, l'Antiope, la Joconde, Marie la Sanguinaire, la Muse de Cortone, la Junon de la villa Ludovici ou Mademoiselle de Lespinasse, Mademoiselle Aïssé, ou Poppée, femme de Néron;—mais telle grisette de Paris, telle jeune fille de salon, telle tête de Burne Jones, telle parisienne de Nittis, etc. . . . —nous fera seule sangloter, nous remuera jusqu'au tréfond de nos entrailles, parce qu'elles sont les sœurs immédiates de notre éphémère, et cela avec son allure d'aujourd'hui, sa coiffure, sa toilette, son regard moderne.[56]

It was a question of the flash of recognition, identification, and, therefore, sympathy. Manet, like Laforgue, captured moments—frequently moments of non-communication between people, as in Laforgue's imagined dialogues, or moments when figures withdrew from the surface world and lost themselves in private meditation. The next minute they would return to the ordinary business of living, but the pause captured by the painting parallels the interior monologues of Laforgue's narrators, and similarly represents the inner level of reverie in an essentially 'modern' and psychological manner.

If the 'modern' interest of Laforgue's subjects in the *Derniers Vers* matched those of the Impressionists, so did his treatment of them, for these paintings were initially criticized for their refusal to take seriously the traditional demands of form, for the haphazard placing of figures in space, and for the unco-ordinated gazes of these figures. In *Le Siècle* of 11 June 1869 Jules Castagnary claimed that: 'Rien d'arbitraire et rien de superflu, telle est la loi de toute composition artistique'; 'comme les personnages dans une comédie, il faut que dans un tableau chaque figure soit à son plan, remplisse son rôle et concoure ainsi à l'expression

[55] *MP* 130. [56] Ibid. 161.

de l'idée générale.' This was precisely what he did not find in Manet's paintings. In a parallel way, Maurice Grammont, in his *Petit Traité de versification française* (1908), refused to entertain the idea that free verse could be considered seriously as poetry,[57] and this also was to be expected. It fulfilled no expectations of regularity, since neither rhyme nor syllabic length could be anticipated: it similarly attacked hallowed traditions. Its harmonies, like those of Impressionist painting, were variable, irrational, and obscure.

The Impressionists' aim of achieving a carefully contrived irregularity was, like Laforgue's, often reached only at a late stage in composition. The most original elements in Manet's paintings have been shown frequently to be late additions; the cat or the choker in the *Olympia* painting, for example, which was originally much closer to Titian's *Venus of Urbino* than it is in its final form, as was *Le Déjeuner sur l'herbe* to the Giorgione-Titian *Concert champêtre.* And, like Laforgue, Manet gloried in originality and 'modernity': 'ils n'ont cessé de me dire que j'étais inégal'; 'ils ne pourraient rien dire de plus élogieux', he insisted. 'Cela a toujours été mon ambition de ne pas demeurer égal à moi-même, de ne pas refaire, le lendemain, ce que j'avais fait la veille, de m'inspirer constamment d'un aspect nouveau, de chercher à faire entendre une note nouvelle.'[58]

If the links between the visual arts and poetry are most clearly seen in the case of Impressionism, Laforgue's poetry nevertheless reflects his many artistic enthusiasms: for Watteau and Gustave Moreau, for the Pre-Raphaelites, and also his more recent German admirations, for Adolf Menzel, like Laforgue—and Huysmans—a 'poet' of factory chimneys, and for Arnold Böcklin, whose bizarre and anachronistic mythological figures influenced not only Laforgue's verse, but also, and more obviously, his *Moralités légendaires.* His aim in art criticism was to convey the sensation of the 'world' created by a particular painter, to write 'une série d'études où, par une accumulation de mots (*sens* et *sonorités*) choisis, de faits, de sentiments dans la gamme d'un peintre, je *donnerai* la *sensation* du monde créé par ce peintre.'[59] He belongs quite clearly to that group of poets, of which Gautier and Baudelaire are the most prominent, who see their task as analogous to, and intimately connected with, that of the painter, in that both create a world from their imaginative and subjective transformation of a changing reality. It

57 M. Grammont, *Petit Traité de versification française* (Paris, 1965), 146-7.
58 A. Proust, 'L'Art d'Édouard Manet', *Le Studio*, 15 (Jan. 1901), 76.
59 *AH* 687 (Laforgue's emphases).

is hardly surprising that Laforgue's poetic aims and methods should match in particular those of the Impressionists, who similarly discovered what might be termed a rhetoric of immediacy.

Laforgue's completion of five short stories,[60] later to be published with one addition[61] as the *Moralités légendaires*, no doubt also had an influence on his poetic manner, since it must have established him in his own eyes as a writer capable of sustained narrative exposition. His earlier excursions into prose had been either brief and unremarkable or had remained in fragmentary form. A connection can be traced between the successful writing of these *Moralités* and Laforgue's move to the longer poem. The *Derniers Vers* are all significantly longer than any of the *Fleurs*. They breathe a greater poetic confidence and expansiveness, which in turn lend the poems greater substance and, ultimately, greater subtlety.

Laforgue's explicit aim in the *Moralités* was to write short stories that were radically different from those of either Maupassant or Villiers de l'Isle-Adam.[62] Given his constant desire for originality, this was hardly surprising. By Pound's time the short story had, in his opinion, 'become vapid, because sixty thousand story writers [had] all set themselves to imitating De Maupassant'. The finely wrought parodic element in the *Moralités*, however, implied 'definitely that certain things in prose were at an end'.[63] Pound draws a parallel between Laforgue's satire in the *Moralités* and Flaubert's in *Bouvard et Pécuchet*, 'if one may compare the flight of a butterfly with the progress of an ox'.[64] What Laforgue also did in the *Moralités*, ironically, was to bring the short story closer to poetry. He speaks of his stories as if they were a volume of poems: 'Je veux . . . faire de mon volume de nouvelles quelque chose de plus qu'un médiocre bouquet de fleurs disparates. Ce sera de l'Art.'[65] Laforgue can scarcely have used the word 'fleur', a recognized metaphor for poem, without intending the parallel to be drawn. It is therefore legitimate to examine the characteristics that the short stories share with Laforgue's verse, and the light they shed on it.

If we consider first their narrative mode and their presentation of character, we see that Laforgue's narrator clearly directs his observations to his reader, keeping himself well in view, so that his sensibility remains the final impression of the stories, as it is of the poems. He

60 'Lohengrin', 'Salomé', 'Hamlet', 'Le Miracle des roses', 'Persée et Andromède'.

61 'Pan et la Syrinx'.

62 *LA* 72.

63 E. Pound, *Literary Essays* (London, 1954), 282.

64 Ibid. 283.

65 *OC*, v. 142.

establishes an early familiarity with the reader, and the style of his stories, as of his verse, frequently has the intimate character of speech and a deceptive appearance of spontaneity. The narrative voice comments, enumerates, questions, and is both unfailingly courteous and at the same time elusively flippant. The techniques vary, as can be seen from the following examples from 'Persée et Andromède', the *moralité* that has been most often linked with the *Derniers Vers*, partly because Leah Lee seems to be the inspiration behind both works:

Elle a poussé ainsi, vous dis-je . . .[66]

Mais où va-t-elle ainsi, ô puberté, puberté! . . .[67]

Pauvre Andromède, on voit qu'elle ne sait pas où prendre son être pour l'exorciser . . .[68]

Ah! elle s'étirera et gémira jusques à quand?[69]

These brief examples illustrate as many narrative modes: direct intervention, exclamation, interpretative comment, interrogation. Narrative intrusions can frequently, as in Laforgue's verse, consist of a single word. For example, Andromède is described as '*irréprochablement* nue'.[70] But the possibility of reproach has its existence in a mind external to the characters, since there is no indication that Andromède is aware of her nudity, and certainly none that either she or the dragon considered it a matter for self-reproach. It belongs to a world that the narrator and reader share.

The short stories offer, therefore, a self-conscious dialogue with the reader, the interior monologue of two or more voices of the self (here the narrator and the dragon, since the dragon is clearly a persona of the poet, as, of course, are Hamlet and Lohengrin), and the interior monologue of the other main character (here Andromède, and always a girl)—that is, precisely the elements that we have already found in many of the *Fleurs* poems, and those which will be given more expanded poetic treatment in the *Derniers Vers*. As Alain-Fournier remarked: 'Il [Laforgue] est à la fois l'auteur et le personnage et le lecteur de son livre . . . *Ça n'est plus du roman*, c'est autre chose.'[71]

If the narrator had already been developed with subtlety and complexity in the *Fleurs*, the 'jeune fille' had been given more summary treatment. The various heroines of the *Moralités* are much more detailed and sympathetic creations, and it is arguable that the 'fiancée' of the

66 *ML* 256.
67 Ibid.
68 Ibid. 260.
69 Ibid. 263.
70 Ibid. 255 (my emphasis).
71 Jacques Rivière and Alain-Fournier, *Correspondance*, i (Paris, 1926), 33.

Derniers Vers, with her strong emotional appeal, owes a great deal to these carefully worked portrayals of femininity. Andromède has been linked by all critics with Leah Lee, and the change in Laforgue's writing doubtless reflects his love. We have a heroine intelligent enough to see through the fake hero, Persée, and to bring about her own salvation by her love for the dragon, so that the story of Perseus and Andromeda becomes that of Beauty and Beast, and offers us Laforgue's only happy ending. Andromède, bored, restless, lonely on her island with only the dragon for company, awaiting a transformation in her destiny that is slow in coming, is the epitome of the feminine situation, as well as being the representation of the 'âme du jour'.[72] Not only is she convincingly human; she is portrayed with a comprehension and a tenderness that are an elaboration on anything we have seen in the *Fleurs*, and which remind us of these qualities in the *Derniers Vers*, in a poem such as 'Dimanches IV', for example.

The *Moralités* consist of two contrasting elements, as must any retelling of myth: a modern and 'realistic' element, and a legendary and imaginary element. Laforgue called the stories 'de vieux canevas brodés d'âmes à la mode'.[73] The later poems offer a parallel contrast between the immediate and modern—conveyed by the main narrative of the poem—and the literary and mythical—conveyed by metaphor and intertextuality. The roles may be reversed, but the essential components are the same, and it is therefore not surprising that some of the techniques of the *Moralités* should find themselves reflected in the *Derniers Vers*. There are many analogies between the 'poetic' passages in the *Moralités* and aspects of the writing of the *Derniers Vers*.

If we continue to take as our main examples 'Persée et Andromède' and 'L'Hiver qui vient', we see that, like the poem, the short story evokes an atmosphere from the start. A poetic mood is immediately set by the imprecise but suggestive details given of the island where Andromède and the dragon live: 'O Patrie monotone et imméritée! . . . L'île seule, en jaunes grises dunes; sous des ciels migrateurs; et puis partout la mer bornant la vue, les cris et l'espérance et la mélancolie.' The interaction between physical and emotional, concrete and abstract, is present in the first sentences of the story. Images support the poetic mood, and frequently have both the delicate precision and the humorous unexpectedness of those in Laforgue's verse; for example, the following description of a brief rainbow: 'et c'est sur le dos des vagues la

[72] *LA* 189. [73] Ibid. 188.

caresse d'un arc-en-ciel comme une riche dorade qui a monté un instant et aussitôt replonge, stupidement méfiante'.[74] The technical devices of poetry abound: the rhetorical question, the theatrical aside, apostrophe, parenthesis, enumeration, and that particular use of repetition that consists of the recapitulation of a word as a basis for its further development, which Laforgue used so frequently in the *Moralités* that his first publishers, perhaps considering the repetition an oversight, frequently cut it out: 'Andromède reste là . . . hébétée devant *l'horizon*, *l'horizon* magique dont elle n'a pas voulu . . .'[75] 'Mais sa [the sea's] plainte ne couvre pas *les petits gémissements*, *les petits gémissements* aigus et rauques d'Andromède, qui . . . scrute sans y penser le mécanisme *des flots*, *des flots* naissant et mourant à perte de vue.'[76]

This technique appears in more complex guises in 'L'Hiver qui vient': the similarity can be seen, however, in two simple examples:

Sur une litière *de jaunes genêts*
De jaunes genêts d'automne.

Les sous-bois ne sont plus qu'un fumier de *feuilles* mortes;
Feuilles, folioles, qu'un bon vent vous emporte . . .

The fact that, as Laforgue wrote the more poetic passages in 'Persée et Andromède', he was moving towards the rhythms and sonorities that he would employ in his free verse is most clear if we transpose two short passages into free verse:

sa chère falaise où la nuit descend,
la nuit sérieuse,
oh! sérieuse pour la vie!
si sérieuse et insaisissable . . .[77]

Où va-t-elle ainsi, ô puberté, puberté!
par le vent et les dunes,
avec ces abois de blessée?[78]

In the second example we can observe even the deceptively casual 'rhyming' of free verse. In both we sense the rhythmical phrasing that is central to this form. 'There are highly wrought passages of prose in the *Moralités* which, given a nudge . . . break into *vers libre*,'[79] writes David Arkell, and by that, ironically, one means that they naturally tend to divide into the balanced binary and ternary subdivisions of regular verse, so that the prose reader experiences the sensation of inhabiting a

74 *ML* 250. 75 Ibid. 274. 76 Ibid. 250. 77 Ibid. 273.
78 Ibid. 255. 79 D. Arkell, *Looking for Laforgue* (Manchester, 1979), 217.

poetic world. In his first *moralité* Laforgue had found the writing of prose a painful process: 'Qu'il est plus facile de tailler des strophes que d'établir de la prose!',[80] he confessed to Charles Henry. But it was a craft he learned as he composed his increasingly 'poetic' stories. In 'L'Hiver qui vient' he moved into free verse with a simultaneous sense of release from the tyrannies of rhyme and the syllabic count, and with the ease and assurance gained in part from the careful shaping of prose sentences into rhythmically balanced units.

THE CREATION OF THE FIRST FREE-VERSE POEM

This was, then, in brief outline, the climate—both external and personal—in which Laforgue moved towards a new conception of what poetry might be. For we cannot doubt that this was a dramatic turning-point in the history of French poetry, which, after the introduction of free verse, could never be remotely the same again. If a later poet, such as Valéry, selected a regular metre for his poem, this now implied a rejection of free verse. If a poet chose to write in free verse, he could do so using a range of subtle alliances with regular metre that his reader was intended to recognize. A free-verse poet who included an alexandrine couplet or an octosyllabic quatrain in his poem now meant something by this inclusion, over and above what was conveyed by the metre itself. Insertions in regular metre took on the status of signs; they had become quotations from another world, and had thus acquired overtones of traditionalism, archaism, or parody. And when a free-verse poet devised novel rhythmical variations or harmonies, these also were enjoyed as deviations from a known norm. No such choices were innocent, and every choice was an implicit statement of intent. Regular metre, with its long and intricate history, stood firmly behind free verse, and it would be very far from the truth to see free verse merely in terms of liberation.

In the early twentieth century young avant-garde poets such as Apollinaire and Max Jacob criticized Laforgue. He was thought not to have been bold enough. Among other things, he was accused of having gone only some of the way towards the liberation of verse: he used so many rhymes, so many octosyllables and alexandrines. The reverse of this accusation would have been more accurate. Laforgue recognized from the start, like Éluard later, that free verse existed in relation to regular verse, that this was unavoidable, and that it was a possible source of richness.

[80] *OC*, v. 128.

If the move to free verse, especially among younger poets, became something of a stampede, the revolution that had taken place was not quickly understood. Mallarmé, as we have seen, so accurate in his description of what free verse actually was,[81] nevertheless termed it 'une heureuse trouvaille',[82] a phrase that might be applied to something as particularized as the 'impair'. (Gérard Genette, with the advantage of hindsight, more correctly calls it 'une mutation . . . profonde'.[83]) Mallarmé's compliment to Laforgue, that he 'nous initia au charme certain du vers faux',[84] also gives the innovation less than its due, while limiting it severely, and it is clear that he really considered traditional verse and prose poetry—the forms he himself used, with the remarkable exception of 'Un coup de dés . . .'— to be the main alternative choices facing the poet. He, of course, considered the 'liberated' alexandrine to be a major innovation, writing of 'L'Après-midi d'un faune': 'J'y essayais, en effet, de mettre, à côté de l'alexandrin dans toute sa tenue, une sorte de jeu courant pianoté autour, comme qui dirait d'un accompagnement musical fait par le poète lui-même et ne permettant au vers officiel de sortir que dans les grandes occasions.'[85] Laforgue thought quite differently. He chose free verse as opposed to regular metre, writing to Kahn emphatically: 'Je ne ferai jamais plus de vers qu'ainsi', and 'j'aurai un volume ainsi en arrivant à Paris. Je ne fais que ça.'[86] And there is no evidence that he wrote again in any other form. Debauve makes the interesting, but in my view not entirely convincing, suggestion that he might have relented to some extent with regard to his poems in *vers libéré* had he lived longer.[87] Clearly, the twelve poems in free verse were too few to make up a volume. It is possible, he suggests, that, after his marriage, when he was already seriously ill, he considered making more use of the *Fleurs*, and, of course, had he added some poems in *vers libéré* to the *Derniers Vers*, he would have ended up with a hybrid volume like Kahn's *Les Palais nomades*, published in 1887. We are not in a position to judge with certainty, since Laforgue's literary correspondence ceased entirely once he was in Paris. My view is that, being more wholeheartedly committed to free verse at this point than Kahn was, he would not have departed from this form. He might have expanded and reshaped poems from the *Fleurs* (as in III, IV, V, VII, VIII, IX, and X), or written new ones (as in XI and XII).

[81] See above, n. 13.
[82] Mallarmé, *Œuvres complètes*, 644.
[83] G. Genette, *Figures* II (Paris, 1969), 124.
[84] Mallarmé, *Œuvres complètes*, 362.
[85] Ibid. 870.
[86] *LA* 194.
[87] J.-L. Debauve, *Laforgue en son temps* (Neuchâtel, 1972), 57.

The evolution of his first free-verse poem, 'L'Hiver qui vient', can be traced with confidence, since three significantly different versions of it survive. They are the early text published by Kahn in *La Vogue* in September 1886, a later undated and undatable manuscript on which some changes have been made in a neat and careful hand,[88] and the text published by Édouard Dujardin after Laforgue's death, containing his latest alterations. The major changes are between the *La Vogue* text and the manuscript, and they are mainly augmentative in character. The poem becomes considerably enriched by two kinds of addition. The first is unexpected, because it consists of the details of the external world, and provides the concrete frame, as it were, of the poem. But if one remembers that the poem is essentially the re-creation of a stream of consciousness, it is perhaps not so surprising that what apparently causes it, but is in reality the background to it, should be developed in its later versions. The 'cheminées d'usines' (which certainly *appear* intrinsic to the poem), the 'journaux', the 'statistiques sanitaires', the 'spectacles agricoles', and even the pianos are not in the *La Vogue* text, and appear in the poem only in the manuscript version. The inclusion of such banal elements alters (as Barbara Johnson argues of Baudelaire's famous 'batterie de cuisine'[89]) not so much our view of the objects themselves, as our perception of the nature of the 'poetic'.

The other kind of addition, a more predictable one, links the poem with art and myth, and so widens its horizons. The monologue acquires richness and depth by these allusions, typically brief, which invite a smile while they offer a spark of illumination. So the 'ornières' (with their Rimbaldian echoes) rise 'en don quichottesques rails', evoking momentarily a delightful world of naïve idealisms, or the picnic baskets conjure up the delicate and elegant idylls painted by Watteau ('Tous les paniers Watteau sous les marronniers'), or the mention of Red Riding Hood transports us to a world of fable. All these references enter the poem only in the manuscript text. (The only reference of this kind present in the *La Vogue* text is a more conventionally classical one, to the river Pactolus, which, like the sun, turns everything to gold.) They help to convey the multilayered quality of the mind, for, as outer landscapes inform the inner 'état d'âme', so the inner landscape surfaces in the images offered by art. They are all more deceiving than they appear. Don Quixote is a tragic as well as a delightfully comic figure; Watteau's idylls do not celebrate anything more elevated than sensual

88 The MS is in the Bibliothèque Jacques Doucet.

89 B. Johnson, *Défigurations du langage poétique* (Paris, 1979), 104.

pleasure; and Red Riding Hood was eaten by the wolf. The poem gains complexity by their inclusion, becoming wittier and more poignant simultaneously.

Laforgue also made diminutive changes, chiefly removing anaphora of too simple a kind.[90] The most interesting of these changes is the play with the poem's ending. In the first line of the *La Vogue* text quoted below he used irony to undercut his wish to portray the season of death:

> Nul n'en rendra raison,
> Tous les ans, tous les ans,
> J'essaierai en chœur d'en chanter la note.

In the manuscript version he moved this irony to the end of the poem.

> Tous les ans, tous les ans,
> J'essaierai en chœur d'en donner la note,
> Pour mes compatriotes,
> Mais qu'on ne me demande pas la raison! . . .

But the last two lines are crossed out, and they are not present in the final text. He was clearly still tempted at this late stage to adopt such self-mocking rhetoric—an echo of his former self, recalling the manuscript dedication of the first poem of the *Fleurs* 'aux français de demain'. The removal of these lines allows the poem to end more affirmatively, with the poet's intention to evoke every autumn the season that he had always found particularly compelling, and which he described in 'Persée et Andromède' as 'esthétique par excellence'.[91]

Of course, we also find substitutive changes, and these reflect a movement towards greater refinement of expression. The following two examples show subtly heightened emotion. In the first, the inclusion of 'loin' in the manuscript version of the description of the soldiers ('Ou les sommiers des ambulances | Pour les soldats *loin* de la France') adds a sense of exile to the existing awareness of their wounds.[92] In the second example intentionality is replaced by anxious enquiry, as 'Vous serez mes seules amours' becomes, after the manuscript stage, 'Serez-vous pas mes seules amours?' Cliché and sentimentality are removed:

90 (1) 'On ne peut plus s'asseoir, tous les bancs sont mouillés': repetition of l. 7, between ll. 40 and 41, crossed out on manuscript. (2) 'O Soleils des soirs tout orangés, aux hallalis, | Aux orations des royaux cors de chasse | Où m'êtes vous ensevelis?': echo of ll. 16–18, between ll. 53 and 54, removed after manuscript.

91 'Mais comme rien n'est plus chatouilleux aux organismes supérieurs que se sentir mourir tout en sachant qu'il n'en sera rien, le crépuscule et l'automne, le drame du soleil et de la mort sont esthétiques par excellence.' (*ML* 261.)

92 The line does not exist in the *La Vogue* text.

thus 'perles d'eau' become 'gouttes d'eau';[93] and the gardens that were first 'pauvres' (*La Vogue*), then 'souffreteux' (manuscript), settle for the understated 'modestes'.

More interesting, from the angle of free verse, is the division of the long Whitmanesque line, retained at the manuscript stage, but divided later. Two examples will illustrate the point (in each case, the manuscript version precedes the final version):

Sur une litière de jaunes genêts d'automne

Sur une litière de jaunes genêts
De jaunes genêts d'automne.

Oh! dans les bruines, toutes mes cheminées d'usines . . .

Oh! dans les bruines, toutes mes cheminées! . . .
D'usines . . .

The first introduces musicality; the second gives 'usines' prominence. The bathos of the unexpectedly short line in this second example is a feature of Laforgue's free verse. The most striking incidence of it in this poem is the addition[94] of the words 'Des spectacles agricoles' to the splendidly bombastic line that precedes it:

Soleils plénipotentiaires des travaux en blonds Pactoles
Des spectacles agricoles.

Laforgue chiefly divides lines, however, for more obviously lyrical effects. The poems after 'L'Hiver qui vient' were, on average, composed of shorter lines of more even lengths, as Laforgue gave increasing importance to the creation of rhythm.

The textual changes to 'L'Hiver qui vient' frequently provide new rhymes, internal rhymes and assonances. When Laforgue introduces the delightful Don Quixote allusion, he is at the same time creating a 'rime alternée' with 'bercails'. The Watteau reference results in a 'rime plate' with the preceding line. When he introduces the image of the ocean of roofs—'Devant l'océan de toitures des faubourgs'—he is creating a rhyme for 'Serez-vous pas mes seules amours?' When he changes:

Que l'autan, que l'autan
Effiloche les savates du temps

to the witty:

93 Before the manuscript text.

94 Between the *La Vogue* and the manuscript text.

Que l'autan, que l'autan
Effiloche les savates que le temps se tricote!

he is not only creating a rhyme for 'la planète falote'; he is also, by means of alliteration and assonance, heightening the sound patterning of the two lines. And this is generally the case. Laforgue's free verse would seem to illustrate Whitman's belief that 'free growth' progresses inevitably to form: 'The rhyme and uniformity of perfect poems show the free growth of metrical laws, and bud from them as unerringly and loosely as lilacs and roses on a bush, and take shapes as compact as the shapes of chestnuts and oranges, and melons and pears, and shed the perfume impalpable to form.'[95]

As one examines this increasingly delicate phonetic and rhythmical progression, it is difficult to accept Kahn's analysis of Laforgue's reasons for moving to free verse as the whole truth: 'Dans un affranchissement du vers, je cherchais une musique plus complexe, et Laforgue s'inquiétait d'un mode de donner la sensation même, la vérité plus stricte, plus lacée, sans chevilles aucunes, avec le plus d'acuité possible et le plus d'accent personnel, comme parlé.'[96] This is a very good description of Laforgue's psychological quest, which is not in doubt, and Kahn defines accurately the two main areas of innovation—the psychological and the musical—but what, one may ask, can the psychological be without the musical, or the musical without the psychological? Kahn is talking about emphases, but, even so, Laforgue's many references to rhythm and musicality in his letters to him make this a puzzling statement. No doubt, in the early 1880s Kahn did awaken or develop Laforgue's interest in the incantatory quality of poetry, and in the spring of 1882 Laforgue wrote to Charles Henry: 'je deviens (comme forme) kahnesque et mallarméen'.[97] In the same letter, however, he made the centrally important, and by now thoroughly familiar, statement that illustrates beyond all doubt his interest in the musicality of verse, and in which we find musicality and psychology combined, as they surely must always be: 'Je songe à une poésie qui serait de la psychologie dans une forme de rêve, avec des fleurs, du vent, des senteurs. D'inextricables symphonies avec une phrase (un sujet) mélodique, dont le dessin reparaît de temps en temps.'[98] Reflecting Laforgue's poetic aspirations at a time when he still wrote in regular metre, this description seems particularly applicable to free verse as he developed it.

[95] W. Whitman, Preface to the *Leaves of Grass*, in his *Complete Poetry* (London, 1938), 574. [96] Kahn, 'Préface sur le vers libre'. [97] *AH* 757. [98] Ibid.

7

Comparisons and Contrasts: Kahn and Eliot

LAFORGUE AND KAHN

TWO preliminary approaches may help to shed light on the originality and the essential character of the *Derniers Vers*. One is to look at the volume that Kahn was writing in the same year, *Les Palais nomades*, since the friendship was a literary one, and the two poets exchanged ideas and felt that they shared artistic aims. The other is to look at the influence that Laforgue exerted on T. S. Eliot, to examine the two major poems composed under this influence—'The Love Song of J. Alfred Prufrock' (hereafter referred to as 'Prufrock'), and 'Portrait of a Lady'—not so much for what they tell us about Eliot, but, somewhat perversely, for what they tell us about Laforgue.

We will begin with Kahn. Individual poems from *Les Palais nomades* and the *Derniers Vers* were published concurrently in *La Vogue* in 1886. The first free-verse poem actually published was Kahn's 'Intermède IV', printed in the June number of *La Vogue*, thus preceding the publication, though not necessarily the writing, of 'L'Hiver qui vient', which appeared in August, but was sent to Kahn in July. 'Intermède IV' is written in what now seems a very hesitant free verse, little more than a *vers libéré*. Syllabic lengths vary from the expected length by one or two syllables (10, 9, 8, 8; 12, 10, 10, 10, 8; 10, 10, 10, 10, 9, 11, 8). We have to wait until the final section of the volume to find free verse that looks anything like Laforgue's on the page.

The book is divided into eight titled sections. A prose poem precedes each section, and sets out its themes. The poems are then numbered, but not titled. The titles of the sections all have musical associations: 'Thème et variations', 'Mélopées', 'Intermède', 'Voix au parc', 'Chanson de la brève démence', 'Lieds', 'Mémorial', 'Finale'—and, indeed, this is where Kahn thought that the chief difference between him and Laforgue lay ('Dans un affranchissement du vers, je cherchais une musique plus complexe, et Laforgue s'inquiétait d'un mode de donner la sensation même, la vérité plus stricte . . .'[1]). It is a musical

[1] G. Kahn, 'Préface sur le vers libre', in *Premiers Poèmes* (Paris, 1897).

influence that is divided between classical and modern, oriental and northern. The melopoeia is allied to the German *Lied.* The whole volume, as its title suggests, has an exotic flavour, which it owes partly to a strong Parnassian influence as well as to a number of echoes of Nerval (the epigraph to the volume is the first couplet of Nerval's sonnet 'Artémis'). To the Nerval influence, for this is undoubtedly a derivative volume, is added a tone of decadent indulgence with an intimate, Verlainian quality.

The range of styles in Kahn's volume is wide, from prose poetry to the classical alexandrine. Kahn's use of devices that he probably owed to Verlaine is more obvious than Laforgue's: particularly archaisms ('Les essaims vont dormant sous les arbres de rêve': 'Intermède' I), and deviant grammar ('Es-tu debout pensant s'exiler les allantes': 'Intermède I'). Kahn and Laforgue employ many of the same devices: anaphora and apostrophe (both frequently used), ellipsis, the one-line stanza, shifts of narrative stance and of register. Kahn also occasionally begins lines in the lower case, thus increasing the sense of fluidity and informality of the poem, a technique that Laforgue never uses. He uses phonetic patterning more insistently and less humorously than Laforgue. His style of writing is well illustrated by his first published free-verse poem, 'Intermède IV'.

> Timbres oubliés, Timbres morts perdus,
> Pas d'une autre glissant à la rue,
> Chansons d'amour et vols de grues
>   Dans d'improbables firmaments,
>
> Les futurs sont à vous, puisque le vent emporte
> Vers des cieux, et des lunes, et des flores
> Vos petits frissons que nul ne peut clore
> Votre âme a glissé sous les lourdes portes
>   Vers d'imaginaires Lahores.
>
> Timbres oubliés des charmants jardins,
> Timbres argentins des Thulés lointains,
> Timbres violets des voix consolantes
> Épandant graves les bénédictions,
> Timbres bleus des péris aux féeries,
> Timbres d'or des mongoles orfèvreries
>   Et vieil or des vieilles nations! . . .

The following characteristics stand out: while the stanzas are of irregular length, a sense of form is conveyed by the progression to the longer stanza; similarly, the variable syllabic lengths are balanced by emphatic phonetic patterning ('péris aux féeries', 'vieil or des vieilles nations') and by insistent anaphora ('timbres' in the last stanza is repeated in five of its seven lines). Verlaine's influence is seen in the poem's deliberately childlike simplicity, in echoes of Verlaine's preferred words ('petits frissons'), and in the adverbial use of the adjective, which Kahn frequently employs ('Épandant graves les bénédictions'). Ambiguity is found throughout: in the synaesthesia of the poem, in the lack of precision of subject and object ('une autre' is referred to; an unspecified 'vous' is addressed), in the loose juxtaposition of elements within the poem, and in the use of plurals ('Chansons d'amour et vols de grues'). Evocation is clearly the aim of this poem, which ends with an exclamation mark followed by suspension points. Why, then, does it leave the reader with a sense of vagueness but little more?

On the face of it, Laforgue's themes and Kahn's are very close: the ephemeral, the internal world of reverie, the gap between the ideal and the real, the clash between the sexes, the idolatry of woman, the search for an acceptable *modus vivendi*, even the attempt to convey the multiple voices of the self. But in this early volume Kahn does not succeed in 'transmuting ideas into sensations', in Eliot's phrase. What was needed was, in the words Kahn used to describe Laforgue's aims, the urge to 'donner la sensation même, la vérité plus stricte, plus lacée . . . avec le plus d'acuité possible'.[2] But, in order to do this, the poet must be obsessed with the 'sensation' and the 'vérité', which he hopes to render in language. He must see the 'précis' as a necessary adjunct to the 'indécis'.[3] Eliot's description of the poetry of William Morris is curiously applicable to that of *Les Palais nomades*: 'the verses of Morris, which are nothing if not an attempt to suggest, really suggest nothing; and we are inclined to infer that the suggestiveness is the aura around a bright clear centre, that you cannot have the aura alone.'[4] The tendency to reach out for the aura rather than for the centre continued in French poetry, and some ten years later Valéry attacked contemporary poets in his *Cahiers*, speaking of their 'impuissance, leur vanité, leur enfantillage, et leur grandiose, dégoûtante répugnance à voir clairement ce qui est'.[5]

After Laforgue's death, as is well known, Kahn claimed to be the inventor of free verse and to have led Laforgue towards it. The prob-

[2] Ibid.

[3] 'Rien de plus cher que la chanson grise | Où l'Indécis au Précis se joint.' (Paul Verlaine, 'Art poétique', from *Jadis et naguère*.)

[4] T. S. Eliot, 'Andrew Marvell', in *Selected Essays* (London, 1932), 300.

[5] P. Valéry, *Cahiers*, ii (Paris, 1974), 1145.

ability is that in 1886, the year of its conscious development, the reverse was the case. What is clear is that Kahn was a distinctly inferior poet, very much at the centre of the currents of contemporary literary innovation, and, later, very interesting as a theorist of free verse, but lacking true creative originality.[6] In his correspondence with Kahn, while praising warmly every element of invention that he finds in his verse, Laforgue also sets him right on some fairly basic points: 'Je ne t'apprendrai pas', he writes to him, probably in December 1884 'que dire à des yeux *clairs miroirs* comme dans Beaudelaire [*sic*] etc., etc. C'est un vilain coup de cymbale connu, obligé, mieux vaut finir sur une quinte!'[7] And one rather delightful passage suggests that he saw very clearly all that was wrong with his friend's poetry: 'quand nous vivoterons ensemble, je saurai bien te pousser à émerger ta tignasse de ces bleus, de ces blonds, de ces finis de glisser, de tous ces cadres albes,—pour gueuler un peu, non plus de cette voix suraiguë de Sixtine, mais de ta voix de la rue Laugier. Afin qu'on dise, comme il est écrit: "Ce Kahn, il faudrait l'incarcérer, il manque de principes et sape les fondements de la vie moderne, etc. . . ."'[8]

LAFORGUE AND ELIOT

About twenty-five years after Laforgue wrote his *Derniers Vers*, the ironic tone that he used still seemed new to the young Eliot, who

[6] For a detailed and illuminating discussion of the origins of free verse, see Clive Scott, *Vers libre* (Oxford, 1990). I interpret the Kahn/Laforgue situation with regard to free verse slightly differently. There can be no doubt that Kahn was technically in advance of Laforgue in 1880, when he claimed that he told Laforgue his views on free verse; he appears to have written no free-verse poems at that time, however. Dujardin's statement, that by March 1886 free verse was for Laforgue 'une chose acquise' (*Les Premiers Poètes du vers libre* (Paris, 1922), 58), has, in my view, been received with too much scepticism. It is probable that Laforgue was already enthusiastic about free verse—not, of course, that he had already written 'L'Hiver qui vient' by that date. Free verse would quite naturally have seemed to him a personal discovery (as it obviously did), because by 1886 he had no doubt forgotten what Kahn may have said to him in 1880. He had gained in literary sophistication during the years when Kahn was removed from the literary scene by his military service, so that, by 1886, as his correspondence suggests, their roles were reversed, and Laforgue was now in advance of his friend.

[7] *LA* 69.

[8] Ibid. 177. Differences between the two friends are highlighted by their contrasting reactions to Tolstoy's *War and Peace*, which they read in 1885 in a French translation. Kahn did not like the work at all (*LA* 141); Laforgue repeatedly expressed admiration for it: 'Je ne connais rien au-dessus de ça. C'est de l'humanité comme il est impossible à un Français de Paris d'en dégager.' (Ibid. 105.) Laforgue was able to look beyond decadentism, whereas Kahn was not.

admitted that his early poetry 'in 1908 and 1909 was directly drawn from the study of Laforgue together with the later Elizabethan drama'.[9] Eliot discovered Arthur Symons's *The Symbolist Movement in Literature* in December 1908, and immersed himself in Laforgue's works chiefly in 1909. Many of Eliot's phrases clearly have their origin in Laforgue. Eliot's capacity for inspired assimilation of the writing of others has often been remarked on, but the fact that Laforgue was a relatively obscure French poet no doubt made it possible for Eliot to adopt phrases and images more liberally than he could have done had he been an English poet or a well-known French one.[10] Had the borrowings been from Donne or Baudelaire, for example, they would have become intertextual signs to his reader. As it was, they were noticed mainly by scholars, and the poems gain nothing by their recognition. Eliot saw in Laforgue a kindred spirit,[11] the processes of whose mind he could understand. He had already observed in Symons's description of him a similar fastidiousness of feeling, high intelligence, and complementary metaphysical interests. From the poetry itself, he discovered a poetic language: 'Laforgue was the first to teach me how to speak, to teach me the poetic possibilities of my own idiom of speech,' he said in a talk in London in 1950.[12] Intent on finding a poetic method suited to the personal confession, he discovered in Laforgue a ready-made *strategy*—the creation of the detached and apparently flippant commentator—which he could employ for his own ends. Using him, therefore, as a kind of springboard, he found in him a source of fresh inspiration. He admired two things in particular in Laforgue: his 'essential quality of transmuting ideas into sensations, of transforming an observation into a state of mind';[13] and his use of irony to express a '*dédoublement* of the personality against which the subject struggles'.[14] Both these qualities

[9] T. S. Eliot, Introduction to *Selected Poems of Ezra Pound* (London, 1928), p. viii.

[10] For some of the many examples of Laforgue's borrowings, see W. Ramsey, *Jules Laforgue and the Ironic Inheritance* (New York, 1953); R. Taupin, *L'Influence du symbolisme français sur la poésie américaine* (Paris, 1929); F. Scarfe, 'Eliot and Nineteenth-Century French Poetry', in G. Martin (ed.), *Eliot in Perspective* (London, 1970); and E. J. Stormon, 'Laforgue and Eliot', *Essays in French Literature*, Nov. 1965: 103–14. Reading Baudelaire and the Symbolists transformed Eliot's early style: almost any two of the poems published in the *Harvard Advocate* 'before' and 'after' would illustrate the point.

[11] Eliot described the kind of influence that could be useful to a writer in these terms: 'a feeling of profound kinship, or rather of a peculiar personal intimacy, with another, probably dead, author' ('Reflections on Contemporary Poetry', *The Egoist*, July 1919).

[12] The talk was given at the Italian Institute, and is reported by H. Howarth in *Notes on Some Figures behind T. S. Eliot* (Boston, 1964), 104.

[13] Eliot, *Selected Essays*, 290.

[14] T. S. Eliot, in *The Criterion*, 12 (1933), 469.

are evident in Laforgue's dramatic monologues as in Eliot's 'Prufrock' and 'Portrait of a Lady'.

Eliot's involvement with Laforgue brought about a thoroughgoing and absorbing reorientation of his basic ideas concerning the kind of poetry that he might write. His neglect of the work of Pound as late as 1910 illustrates this: 'I was introduced to *Personae* and *Exultations* in 1910,' he wrote much later, 'while still an undergraduate at Harvard. The poems did not then excite me any more than did the poetry of Yeats. I was too much engrossed in working out the implications of Laforgue.'[15] One is reminded of Sylvia Plath's confession of willed immersion in the works of others: 'I will imitate until I can feel that I am using what he can teach,'[16] she said of the writings of Frank O'Connor. Examples of such obsession are striking in the visual arts: Matisse before Cézanne's bathers, Francis Bacon before Velázquez's portrait of Pope Innocent X, or Derain 'working out the implications' of Seurat or of Gauguin.

Of course, Eliot's poems of this period differ from Laforgue's in many ways. He was more concerned to establish a 'character' for his narrator, and gives him many more external characteristics. His monologues have something in common with those of his other mentor, Browning, of whom Laforgue would have been less aware.[17] Like his, they set and develop a dramatic situation, even if it is an imaginary one. They imply criticism of a particular attitude—hypersensitive and negative—and of a particular artificial and pretentious society. We can visualize Prufrock and the environment that he inhabits more precisely than we can the narrator of Laforgue's monologues. Much later, in 1953, showing remarkable consistency of attitude, Eliot claimed that pure dramatic monologue could not 'create a character'.[18]

[15] T. S. Eliot, 'On a Recent Piece of Criticism', *Purpose*, Apr.-June 1938.

[16] *The Times Literary Supplement*, 21 June 1991: 19.

[17] In 1929 Kahn reported a detail of considerable interest: that he remembered two books in particular on Laforgue's table in his apartment in Berlin, a copy of Tennyson, and one of Browning (*Les Nouvelles littéraires*, Jan. 1929). The fact that Laforgue never mentions Browning in his letters *may* merely be due to his awareness of his correspondents' ignorance of this poet. Browning's poetry may have helped him to see the dramatic monologue as a legitimate form. Laforgue's monologues are so different in character from Browning's, however, that we probably would not want to take the possible influence further.

A recent study by Elizabeth A. Howe, *Stages of Self: The Dramatic Monologues of Laforgue, Mallarmé and Valéry* (Athens, Oh., 1990), offers an illuminating account of the dramatic monologue, tracing the form from Browning, through Laforgue and Mallarmé, to Valéry and Eliot.

[18] T. S. Eliot, 'The Three Voices of Poetry', in *On Poetry and Poets* (London, 1957), 95.

Obviously, Laforgue was aware that a convincing narrator needed to be seen from the outside at least some of the time, and he used this technique effectively for purposes of ironic deflation, as in the final stanzas of *Derniers Vers* III and VI. He shared with Eliot an interest in the fictional presentation of character (as is shown in his *Moralités légendaires*), which Eliot owed in part to his admiration for Henry James. But in the *Derniers Vers*—and even in the *Moralités*—he was not so much concerned with the credibility or autonomy of his narrators, as with their disjointed inner worlds. Since his poetic aim was to render in language 'des bouts de rêverie sans suite',[19] 'de la psychologie dans une forme de rêve',[20] he inevitably presented snatches of lived experience rather than the being who experienced them.

Prufrock may not represent Eliot himself—for one thing he is much older—but, nevertheless, his malaise emerges vividly as a personal disability. Laforgue's more depersonalized narrator describes more representative youthful dreams and aspirations. It would be futile to apply moral criteria to him, as one is tempted to do to Eliot's narrator, whose attitude, one may feel, lacks warmth, conceals cruelty. In Laforgue we are dealing with the fantasies of the mind, quite literally 'such stuff as dreams are made on'. He explores sadism more openly than Eliot, but it is the sadism inherent in human nature in general, and it is balanced by a tenderness not found in Eliot's more particularized narrators.

Eliot exposes with superb clarity the paradoxes and inconsistencies of human desires and behaviour, the 'dissociation of sensibility' in the context of the potentially dramatic incident. His monologues are not, like Laforgue's, the sensitive listening to the inner voice, so much as the presentation of a powerful inner vision in conflict with the real world. Such is Eliot's desire for poetic synthesis that order is clearly discernible behind the apparent disorder of 'Prufrock'. Laforgue pursues his world of reverie in a more relaxed and more genuinely humorous manner. For example, he gives us the imagined—but unrealistic—speech of the girl who is the subject of the dream, a step that introduces a further ironic level and, incidentally, a further degree of psychological realism. Female characters speak in Eliot, but their speech is realistic, in the sense that their words could actually have been uttered. This is sometimes the case with Laforgue, but at other times their speech represents nothing more substantial than subjective wish-fulfilment fantasies.

Laforgue's narrator, then, is 'characterless' when compared to Prufrock or the narrator of 'Portrait of a Lady', and the 'disease' from which he

[19] *AH* 792. [20] Ibid. 757.

suffers is inherent in his youth. This gives him a universal appeal and also a more protean flavour. In the final poem of the *Derniers Vers* he accepts life, if very tentatively, and we have a vivid sense of the possibility of such transitions from one state of consciousness to its apparent opposite. Prufrock is a complex personality, but he lacks such freedom. Were he to 'change', the reality of the monologue would be seriously undermined. The same is true of 'Portrait of a Lady', where the difference in age and sensibility between the two protagonists has to be seen as insuperable. Laforgue's impasses relate to broader metaphysical issues and conflicts; they are set in a more fluid and shifting psychological landscape. Man is here already to a greater extent a 'structure of the fictions he tells about himself'.[21]

Francis Scarfe has pointed out that the theme of both 'Prufrock' and 'Portrait of a Lady' can be found in the opening lines of 'Dimanches III', which, as we have seen, were taken from three separate *Fleurs* poems:[22]

> Bref, j'allais me donner d'un 'Je vous aime'
> Quand je m'avisai non sans peine
> Que d'abord je ne me possédais pas bien moi-même.
>
> (Mon Moi, c'est Galathée aveuglant Pygmalion!
> Impossible de modifier cette situation.)
>
> Ainsi donc, pauvre, pâle et piètre individu
> Qui ne croit à son Moi qu'à ses moments perdus,
> Je vis s'effacer ma fiancée
> Emportée par le cours des choses,
> Telle l'épine voit s'effeuiller,
> Sous prétexte de soir sa meilleure rose.

Eliot's two poems, both, like Laforgue's, reworked from fragments, and, in the case of 'Portrait of a Lady', composed over a relatively long period, can be read as dramatized expansions of these lines, while Laforgue's monologue takes this situation, compressed into eleven lines, as a starting-point, and the poem continues as the free association of the poet's reverie, passing through loneliness, melancholy, sadistic fantasies, romantic aspiration, sober optimism, and protective tenderness to the final note of ironic detachment. The external world promotes the reverie, lends it its support. The most striking example of this is the coach journey, suspended from time in 'Solo de lune', but always the monologue is assisted by the temporary removal of the real world of

21 P. Brooks, *Reading for the Plot* (Oxford, 1984), 277.

22 Scarfe develops this idea in 'Eliot and Nineteenth-Century French Poetry'.

action. In a sense this is true also of Eliot, but within his monologues he developed images of confrontation, whereas Laforgue indulged the solitary tête-à-tête, checked only by the poet's or narrator's irony. He, like Eliot, faced the bleakness of despair, of course, but he saw it in terms of the nature of life in general, and of the suffering of others as well as himself. The self-disgust found in Eliot—or in Baudelaire—is not a prominent feature in Laforgue. On the face of it, his is the more pitiable situation, but Eliot emerges as the more tormented personality.

Despite similarities of technique and a common reliance on irony—used to convey pathos and humour simultaneously—the monologues of Eliot and Laforgue suggest, therefore, very different atmospheres. Laforgue stands between Mallarmé and Valéry, though he resembles neither closely, in a tradition that the Eliot of this period would probably have considered too inward-looking and insufficiently morally oriented. But Eliot learned a great deal from Laforgue: the value of a certain kind of rhythmical free verse in which expansions and contractions of metre, and irregular rhyme-patterning echoed the shifting play of emotions (as he admitted); a certain brand of irony that emphasized the *dédoublement* of the personality, and which opened the doors of the confessional poem to him. Laforgue was responsible for proving to Eliot that it was possible to present in verse an interior monologue—disjointed and often incongruous—which expressed basic human conflicts: that between the outer world of action and the inner world of sensitive reaction; that between the sexes, estranged from one another by an artificial society; the traditional conflict between body and soul, and the 'modern' conflict between the intellect and the emotions that obsessed both poets. The latter led naturally to what Pound termed 'verbalism', defined by Taupin as the 'enlacement de mots, d'expressions et d'idées harmonisées par des rythmes et des refrains'.[23] Taupin quotes three lines from 'Portrait of a Lady' that he considers to be 'bien du Laforgue de la meilleure manière':[24]

> This music is successful with a 'dying fall'
> Now that we talk of dying—
> And should I have the right to smile?

And, as one thinks of examples such as the following, one is bound to agree:

[23] René Taupin, *L'Influence du symbolisme français sur la poésie américaine*, Paris, 1929, 152.
[24] Ibid. 232.

Vraiment je ne songe pas au reste; j'attendrai
Dans l'attendrissement de ma vie faite exprès.
(*Derniers Vers* IX)

Oh! que nous mourrons!
. . . .
O Nature relève-moi le front!
Puisque, tôt ou tard, nous mourrons. . .
(*Derniers Vers* XII)

Laforgue—along with Baudelaire and, to a lesser degree, Corbière—also gave Eliot a model for images of extreme boldness, such as emerge from the subconscious mind and are highly successful in the context of the free association of the interior monologue. Laforgue wrote, for example, of the winter sun: 'Un soleil blanc comme un crachat d'estaminet', which lay on the horizon 'comme une glande arrachée dans un cou'.[25] He may also have been in part responsible for alerting Eliot to the value of casual intertextual allusions in the interior monologue ('L'Hiver qui vient' alludes to Don Quixote, Watteau, Little Red Riding Hood, the river Pactolus; while 'Prufrock' mentions Michelangelo, John the Baptist, Lazarus, Hamlet), as well as for suggesting whole areas of imagery, such as the submarine for the evocation of the subconscious mind. (The urban, the cosmic, the medical, and the scientific, all used by Eliot at this period, are, of course, common also to Baudelaire and to Rimbaud.)

Laforgue's narrators present the problems facing youth and inexperience more broadly and with more normality. Their monologues are less dramatically powerful than Eliot's, but they are more relaxed, more fanciful. They deal, like Eliot's, with the clash between external and internal worlds, but, in their nuanced and varied tones, they belong to a narrator less radically alienated from the external world, a narrator who inhabits his inner world, melancholy though it may be, more comfortably. Differences highlight the characteristics of each poet, but what the two writers share is also important for our purposes. If Eliot is 'modern', as Christopher Ricks claims, in part because he 'repeatedly refuses the satisfactions, the complacencies, of the secure and indubitable',[26] it must be because Laforgue had led the way. And Laforgue might have appreciated the irony that, in these poems, he goes further along this path than his successor.

25 'L'Hiver qui vient'.
26 C. Ricks, *T. S. Eliot and Prejudice* (London, 1988), 14.

A comparison of Laforgue with Kahn and Eliot, two poets much involved in the creation of modern poetic styles, and with claims to innovation along the same lines as Laforgue, is bound to emphasize still further Laforgue's outstanding originality. Kahn may have been moving in a parallel technical direction, but his achievement is crippled by the privileging of musicality over psychology, by his neglect of meaning—or, more exactly, of the level, in Emily Dickinson's words, 'where the meanings are'. Eliot, obviously a great poet, was in fact in these early monologues rather less forward-looking and experimental than Laforgue was. The later *Waste Land*, of course, has a range and power of which the 26-year-old Laforgue would have been incapable. Nor would any of his literary friends and mentors—Kahn and Bourget included—have been capable of offering him the inspired help that Pound gave Eliot.

8

The *Derniers Vers*: 'Les éclairs d'identité entre le sujet et l'objet'

OUTER AND INNER WORLDS (*DERNIERS VERS* I-II)

WITH the exception of *Le Mystère des trois cors*, which is a kind of fable, the poems that make up the *Derniers Vers*, now generally considered to be Laforgue's most remarkable achievement, are a series of meditations or interior monologues that, as is usual in introspective poetry, draw on the external world for the images that bring the inner world to life. The first poem, 'L'Hiver qui vient', is a soulscape in Verlainian style, depicting vividly the seasonal moment that Laforgue found 'esthétique par excellence',[1] the transition from autumn to winter, and this image, with its accompanying falling leaves, rain, wind, and storms, continues as a background to the succeeding poems, which describe a spleen very different in its colouring from Baudelaire's. From his earliest verse, Laforgue had treated natural phenomena as symbols of aspects of human life: the sunset and the moon are obvious examples. (The cold brilliance of the stars to some extent displaces these earlier symbols in this volume.) His late autumn is a development of, and a variation on, the tradition that includes Hugo, Baudelaire, and Verlaine.

His friend Téodor de Wyzewa, writing in the *Revue wagnérienne*, described the Symbolist aim thus, placing the emphasis on the symbolic value acquired by the objects of the real world when re-created by the poet:

> Nous voyons autour de nous des arbres, des animaux, des hommes, et nous les supposons vivants; mais ils ne sont, ainsi perçus, que des ombres vaines, tapissant le décor mobile de notre vision; et ils vivront seulement lorsque l'artiste, dans l'âme privilégiée duquel ils ont une réalité plus intense, leur insufflera cette vie supérieure, les *recréera* devant nous. . . . Mais où l'artiste prendra-t-il les éléments de cette vie supérieure? Il ne les peut prendre nulle part sinon dans notre vie habituelle, dans ce que nous appelons la réalité.[2]

[1] *ML* 261.
[2] Quoted by A. G. Lehmann, in *The Symbolist Aesthetic in France* (Oxford, 1968), 296.

This is clearly what happens to the elements that Laforgue draws from the external world, and he suggests in his art criticism that this must be the artist's aim:

Si insuffisante et drapée de banalité que soit la formule, il n'y en a pas de plus infaillible. Donner l'âme des êtres et des choses telle qu'elle va à votre âme particulière et éphémère d'artiste: l'âme d'une scène historique, ou de rue ou de salon, l'âme d'un portrait, l'âme d'une nature morte d'huîtres, l'âme d'un tronc d'arbre on d'un terrain d'hiver, l'âme d'un jardin de banlieue ou d'une cheminée d'usine, l'âme d'un torse musclé, l'âme des minutes.[3]

But Laforgue was not only concerned with conveying the symbolic essence of elements of the external world: he was aware that elements that were often trivial in themselves had the power to convey the poet's affective inner world—to evoke 'those indefinable and yet precise emotions' of which Yeats speaks. Many of Laforgue's images—personal and subjective, drawn from the surface world of the modern city, neither expanded nor intellectualized—function impressionistically, doing no more than strike a spark between the outer world and the poet's sensibility. 'Telle grisette de Paris . . . nous fera seule sangloter',[4] he had written when describing the appeal to the emotions of Impressionist painting. T. S. Eliot's later description of what he termed the 'objective correlative' (rechristened by Robbe-Grillet the 'objet-signe') is relevant and still illuminating: 'The only way of expressing emotion in the form of art is by finding an 'objective correlative'; in other words a set of objects, a situation, a chain of events which shall be the formula of that *particular* emotion; such that when the external facts, which must terminate in sensory experience, are given, the emotion is immediately evoked.'[5]

In Laforgue's poetry the situations and chains of events are the small-scale dramas that figure repeatedly in the *Fleurs*, exposed in snatches of clipped dialogue between past or would-be lovers, or they are the lonely wanderings of the outsider in alien urban areas—increasingly Berlin rather than Paris—in empty *fin de saison* resorts or in deserted woods—situations rooted in actual experience of the city, or drawn from the court's routine migrations to Koblenz, Baden-Baden, or Babelsberg Castle. The objects are wet park benches, telegraph poles along empty roads, factory chimneys, fountainless squares, shut-up houses, curtained windows, stations, quays, jetties—images stressing exclusion and

[3] *MP* 169. [4] Ibid. 161.
[5] T. S. Eliot, *Selected Essays* (London, 1932), 145.

ephemerality. 'External facts . . . which must terminate in sensory experience' are not merely visual objects, of course; they are also sounds, with their contrasting emotional resonances: hunting-horns—again, despite their extended literary history, combining court experience with childhood memories for Laforgue—church bells—inseparable in his mind from the lonely Sundays when he watched files of immaculate convent girls on their way to Mass—and the sound of piano scales emerging from faceless houses, conjuring up, as he saw it, an imprisonment parallel to the religious one, that of the genteel bourgeois girls of his fantasies. There are warmer touches: the vanishing beauty of nature seen in the yellow broom of 'L'Hiver qui vient', for example, or images of shared winter comfort—lamps, engravings, *petits fours*—which contrast with a more pervasive set of images suggesting the trivial, the sordid, or the diseased—newspapers, dirty linen, spittle outside drinking-dives, medicines.

None of the elements mentioned, some of which recur insistently, is foreign to Laforgue's experience. He conveys the double impression that the translation of the world into poetry depends on this range of precisely observed details, and that poetry alone can give them their full significance. This was how it seemed to Alain-Fournier, who responded to this fusion in Laforgue of external and internal—this synthesis of perceiving subject and object perceived:

> Ce que j'ai cherché tout de suite—avec passion—chez Laforgue et ce que j'ai trouvé, ce sont, par instants, comment dire? des vers, des bribes de phrases, qui étaient l'expression parfaite et poignante de quelque chose. Une vision. Une impression sentie qui m'allait droit au cœur . . . J'avais l'impression à ces trouvailles (comme je voudrais l'avoir pour tout ce que j'écris) que ça n'était pas *écrit* comme le reste, tant c'était précis et senti.[6]

And, although Laforgue would not have claimed genius for himself, when attempting in his notes on Impressionism to explain what constituted it, he described it as this elusive fusion of subject and object: 'l'objet et le sujet sont donc irrémédiablement mouvants, insaisissables et insaisissants. Les éclairs d'identité entre le sujet et l'objet, c'est le propre du génie.'[7] Clearly, the emphasis on 'éclairs d'identité' is inseparable from our view of modern art, with the importance that it attaches to immediacy.

[6] Jacques Rivière and Alain-Fournier, *Correspondance*, i (Paris, 1926), 143–4.
[7] *MP* 141.

Laforgue's inner world, illuminated by such lightning flashes of identification with the external world, has seemed to some critics to be so consistent—the poems so much a set of variations on a central theme—as to warrant reading the *Derniers Vers* as a single poem. Martin Turnell put forward this view in an article in *Scrutiny* in 1936: 'The *Derniers Vers* is a poem of 816 lines which registers the constant shift and change of feeling, the play of feeling within a prevailing mood, that a sensitive person experiences in modern urban conditions.'[8] After this statement he refers to the individual poems as sections. It is clear from his references to *The Waste Land* that Turnell was reading the *Derniers Vers* through that particular work. What these two, essentially dissimilar, works share, their free verse apart, is this sense of a psychological climate laid bare by a network of repeated and interacting motifs. French critics such as P. Reboul and D. Grojnowski have resisted the 'single poem' approach, while M. Collie and J. A. Hiddleston find it attractive. Some of the individual *Derniers Vers*, they suggest, seem thin and insubstantial taken in isolation, but, as parts of a single structure, in which themes and variations create a complex unit, they acquire their full value.[9]

There are difficulties about this approach. There is no suggestion in Laforgue's correspondence that he was planning an organized whole comparable even to *Une saison en enfer*, let alone anything remotely similar to the highly structured *Waste Land*. When he writes to Kahn of individual poems, always described as separate units, his emphasis is on the new free-verse form. 'L'Hiver qui vient' is clearly the poem that introduces the breakthrough. 'La Légende' (as it was then called) 'des trois cors' is sent merely to amuse his friend: 'Je t'envoie "la légende des 3 cors" pour te distraire.'[10] The statement he made about 'L'Hiver qui vient', 'J'aurai un volume ainsi en arrivant à Paris',[11] suggests a volume of individual poems rather than a single extended poem.

8 *Scrutiny*, 5 (1936), 136.

9 This is the argument put forward by Hiddleston: 'Il est vrai que chaque poème peut se suffire à lui-même, mais lu dans l'ensemble des *Derniers Vers* avec les images et rythmes qui reviennent dans des circonstances et une atmosphère différentes, il gagne en profondeur, en résonance et en valeur émotive.' (*Essai sur Laforgue* (Lexington, Ky., 1980), 35.) Collie's argument is similar: 'Even the strongest poems must gain in imaginative coherence by being seen in context' (M. Collie, *Jules Laforgue* (London, 1977), 62), and of *Derniers Vers* XII he writes: 'the poem then ends with lines which, despite the reference to Baudelaire, would be hollow indeed without the substance of twelve poems behind them' (ibid. 64). The discussion is referred to by Hiddleston in *Hidd. LA* 57.

10 *LA* 195.

11 Ibid. 194.

Other evidence related to the intended structure of the volume is supplied by dates of publication. Ten of the twelve *Derniers Vers* were published in pairs in Kahn's journal *La Vogue* between 16 August and 6 December 1886. The eleventh poem was published separately, under Wyzewa's auspices, in the November issue of *La Revue indépendante*, and the twelfth poem was not published until more than a year after Laforgue's death, in December 1888, again in *La Revue indépendante*. Nine of the twelve poems have titles; IX, X, and XII do not. In the version published by *La Vogue*, however, IX, which is the shortest of the *Derniers Vers*, and X were combined and entitled 'Les Amours'.[12] This leaves Laforgue's last unpublished poem as his only untitled one. The use of titles would seem to suggest that Laforgue conceived of the poems as independent units, although, of course, titles can function as section headings, as they do in *The Waste Land* and in *Une saison en enfer*. The method of publication certainly suggests that Laforgue felt that the poems could stand independently of each other.

In his preface to the limited 1890 edition of the *Fleurs* and the *Derniers Vers*, for which Dujardin and Fénéon minutely transcribed variants, Dujardin refers to the *Derniers Vers* as 'douze pièces que nous appelons spécialement, à défaut de titre, *Derniers Vers*' and makes it clear that the twelve poems did not constitute a completed work: 'La série des *Derniers Vers*, que eût épuisé les éléments du livre *Des Fleurs de bonne volonté*, ayant été interrompue par la mort de l'auteur . . .' In his work *Les Premiers Poètes du vers libre* he refers to them similarly as 'les poèmes qui ont été réunis sous le titre de *Derniers Vers*'.[13] I have discovered no evidence to suggest that Laforgue, or anyone else at the time, thought of the *Derniers Vers* as either a completed volume or a single poem. (Dujardin does not say in his 1894 edition of the *Poésies complètes*, as Collie states, that 'he [Dujardin] had Laforgue's authority to publish the *Derniers Vers* as one poem'.[14] He merely says: 'Peu de temps avant sa mort, Laforgue, qui venait de me confier le soin d'éditer son volume de prose, les *Moralités légendaires* (Editions de la *Revue indépendante*) me demandait d'éditer également ses nouveaux vers.'[15])

[12] On the MS in which he separates the two poems Laforgue simply writes 'autre pièce' before what is now the beginning of X. This, again, supports the view that he envisaged these poems as separate units.

[13] E. Dujardin, *Les Premiers Poètes du vers libre* (Paris, 1922), 56.

[14] Jules Laforgue, *Derniers Vers*, ed. M. Collie and J. M. L'Heureux (Toronto, 1965), 57.

[15] Jules Laforgue, *Poésies complètes*, ed. E. Dujardin (Paris, 1894), 'Avis', p. vi. The MS of a series of cards (now in the Bibliothèque Jacques Doucet) that Laforgue wrote in

We must conclude, therefore, that what we have is an incomplete work. The network of echoes that is set up both between the poems and also within each poem is offset by the discontinuity that is often achieved, as we have seen, by an actual 'patchwork' technique of composition. The semantic balance between repetition and difference is paralleled on the metrical level, as we shall see later, when we examine the poems' free verse.

On the thematic level, the unifying factors are familiar from the *Fleurs*. Uncertainty concerning personal identity, and preoccupation with love are central. The patterns of fear and desire are explored in poems that dwell on the contrast between the ideal and the real, the innocence of the dream 'fiancée' and the violence of male sexuality. We have two opposed set of data: on the one hand, repetition, stagnation, inhibition; on the other, passionate impulses, rebellion and despair, aspiration towards understanding, yearning for love, or acute nostalgia for lost love. A complex landscape of emotions is mapped out in poems that are essentially psychological, and which explore the changing patterns and moods of relationships, real or imagined: doubts, hesitations, confidences, reproaches, resentments, jealousies, misunderstandings, regrets, hopes, and the changes brought about by time, memory, or quite simply by the distortions of the imagination.

Centred on Laforgue's exploration of an inner landscape, the poems are technically varied and experimental. The contrast between the first two poems has already been mentioned. 'L'Hiver qui vient' is a powerful soulscape—even, arguably, a 'poème manifeste', as Lawrence Watson has termed it[16]—while 'Le Mystère des trois cors' is a parodic fable that may illustrate in traditional fashion that love has the power to kill, but chiefly pokes fun at the personified hunting-horns, the 'characters' of its sexual triangle. The poems that follow this 'legend' are all meditations or interior monologues, but they vary in tone,

the first half of 1887 to Dujardin make painful reading, concerned as they are only with payments for his writings, which he clearly needed in order to survive. Dujardin was publishing in *La Revue indépendante* six monthly 'chroniques parisiennes' written by Laforgue, for which he was often paid ahead of publication. Dujardin was also arranging for the publication of the *Moralités légendaires*, which did not appear until after Laforgue's death. In Laforgue's circumstances of extreme hardship, it is not surprising that the free-verse poems that he had written for his next volume were not even mentioned. He had, of course, himself paid for the publication of *Les Complaintes* and *L'Imitation de Notre-Dame la lune*, and knew that he would have to pay for his most recent poems to be published, as is clear from a letter of Jan. 1887 to Léon Vanier (J.-L. Debauve, *Laforgue en son temps* (Neuchâtel, 1972), 131).

[16] L. Watson, '"L'Hiver qui vient": Poème manifeste', in *Hidd. LA* 135–53.

breadth, and degree of complexity. We have considered III, VII, and VIII, which create a complex whole with the use of quotations from the *Fleurs*. An especially intricate semblance of 'patchwork' is created in the first and last poems of the volume without recourse to any element from the *Fleurs*.

Whatever their differences, the system of equivalences built up between external and inner worlds is strengthened by a series of allusions to literature, religion, philosophy, art, myth, and folklore. Originating in Laforgue's reaction to the external world and all attempts at some kind of description of it, they are present in the poems as part of the poet's inner world, shaping and colouring his view of the nature of existence. *Hamlet* is, of course, the most prominent. The volume is prefixed by statements from three characters from *Hamlet*: Hamlet's original confession of love to Ophelia (inaccurately quoted by Laforgue), Ophelia's speech expressing her bewilderment at Hamlet's apparently deranged behaviour, and Polonius's too simple conclusion: 'This is the very ecstasy of love.' At the head of the final poem Laforgue placed Hamlet's 'Get thee to a nunnery' speech. In the text of the *Derniers Vers* Hamlet's exclamation 'Alas poor Yorick!' is quoted, and classical heroines who, like Ophelia, are victims of destiny (Antigone and Philomela) are referred to, though Ophelia herself is not. That Laforgue thought of the parallel between himself and Hamlet as he wrote the volume is perhaps most vividly brought home to the reader by the fact that he placed his own initials on the manuscript after Hamlet's confession of his love to Ophelia; which is, as we have seen, the epigraph to the poems:

> I have not art to reckon my groans thine evermore
> Most dear lady, whilst this machine is to him.
>
> (J.L.)

While one of the dominant images of the volume is a religious one—church bells, with their associations of religious services and convent girls—the commonest references are to Greek mythology: to Circe and Cybele, the rivers Pactolus and Lethe, Venus, the Bacchic thyrsus. The poems also convey a sense of absorption in the whole Western cultural tradition by their casual references, implying an easy recognition on the part of the reader, to *Le Roman de la Rose*, Don Quixote, Watteau, or Wagner's Valkyries, for example. Ironic Baudelairean echoes are frequent, and the final poem ends on a rewriting in modern idiom and sensibility of Baudelaire's famous last lines to 'Un voyage à Cythère':

O Nature, donne-moi la force et le courage
De me croire en âge.

Nature, the unconscious force dominating the physical world, replaces Baudelaire's divinity; a simple desire for life his overwhelming sense of evil.

In other cases—that of Léon Dierx, for example—a once well-known model has now become unfamiliar. Watson places Dierx's 'Soir d'octobre'[17] firmly behind 'L'Hiver qui vient', and considers that Laforgue, here and elsewhere, uses parodic echoes of Dierx's themes and images to mark the difference between an earlier style of writing and his new, self-consciously 'modern' lyricism.[18] Dierx's depiction of reality, it is argued, had lost what power it might once have had to interest or convince. By setting his own against it, Laforgue draws his reader towards his version of the 'real'. To some extent, therefore, Laforgue's approaching winter owes its vividness and poignancy to the fact that it is *not* that of Dierx. Dierx is nowadays an almost forgotten writer, but in the late nineteenth century he was both well-known and respected. He was awarded the title of 'prince des poètes' after Mallarmé, supported, it would appear, by Mallarmé himself. Although Laforgue never mentions Dierx in his correspondence, it seems likely that he knew at least some of his poetry, and equally likely that he felt as distanced from it as he did from the solemnity of his own early style. The climax of Dierx's 'Soir d'octobre' is an earnest address to the human soul:

Âme de l'homme, écoute en frémissant comme elle [the church bell]
L'âme immense du monde autour de toi frémir!

Laforgue, by contrast, ends 'L'Hiver qui vient' with a purely literary aspiration, the desire—a suitably tentative one, expressed by a future verb—to capture the atmosphere of late autumn in his poetry:

Tous les ans, tous les ans,
J'essaierai en chœur d'en donner la note.

'En chœur' refers to the artistic attempt at orchestration of his solitary inner experience, but the inwardness of this experience is rendered in the pun with 'cœur'. The distance between outer and inner worlds is acknowledged, while the artistic task is set clearly as the discovery of the illuminating spark between the two.

[17] L. Dierx, 'Soir d'octobre', *Poésies* (Paris, 1972), 146.
[18] Watson, '"L'Hiver qui vient"'.

9

The *Derniers Vers*: 'J'ai voulu faire de la symphonie . . .'

DISCONTINUITY: THE DRAMATIZED AND DEPERSONALIZED MONOLOGUE (*DERNIERS VERS* III-VI)

WE have seen the dramatic character of Eliot's monologues, which make more use of the theatrical elements of plot and character than Laforgue's do. Laforgue's monologues are dramatic chiefly in the collisions caused by their sudden and extreme shifts of narrative stance. In the final stanza of 'Dimanches III', for example, there is a total volte-face. The narrator whose monologue we have been following is addressed in a series of imperatives:

> —Allons, dernier des poètes,
> Toujours enfermé tu te rendras malade!
> Vois, il fait beau temps, tout le monde est dehors,
> Va donc acheter deux sous d'ellébore,
> Ça te fera une petite promenade.

Since hellebore was the traditional cure for madness, little of the preceding monologue would be left standing if the reader were able to trust this pseudo-external voice of reason. But, as in 'Avant-dernier Mot', it is merely one of the many voices of the self, and one particularly given to platitudinous comment. Earlier in the poem we find similar narrative shifts: apparently objective description is abruptly interrupted by an apostrophe to the loved girl that smacks of self-pity:

> Mon corps, ô ma sœur, a bien mal à sa belle âme. . .

Later a reverie about the relations between the sexes is cut short by an urgent appeal that appears, on the surface, more than a little ridiculous:

> En attendant, oh! garde-toi des coups de tête,
> Oh! file ton rouet et prie et reste honnête.

Each of these interruptions in the present tense dramatically alters the perspective. They all also originally belonged to different poems, for, as we have seen, 'Dimanches III' was made up of quotations from six

Fleurs poems to which Laforgue added new elements. An effect approaching that of the twentieth-century stream of consciousness is thus conveyed by a totally artificial method. We are reminded that for 'Prufrock' Eliot also pieced together fragments written over a period of time. Clearly, surrealists would not have considered that the contradictions and discontinuities of the mind needed to be reached in such a complicated and unnatural manner. But this is how an interior monologue with this degree of dramatic complexity in fact originated in French poetry.

Although the monologue makes use of narrative, it takes the reader to an area that lies beyond it, where the various ingredients of the storyline coexist because, quite simply, they are continuously being turned over in the narrator's mind. In III, for example, the time progression is undercut by the analytical tone of the first and last stanzas, which links beginning and end and which suggests the circularity of the patterns of the mind. Nothing approaching a solution has been reached, since the final suggestion of madness is clearly ironic and is further lightened by the reference to La Fontaine's fable about the tortoise and the hare. Although the narrator's moods follow in rapid succession, they are all obviously a permanent element of his consciousness. Since this is so, since Laforgue's narrator is depersonalized (though not necessarily impersonal), the fact that the monologue may be created by this artificial patchwork technique seems immaterial. Each poem is a variation on the theme of the same internal landscape. This is a reason for the repeated echoes that can be traced between the poems, and which have given rise, as we have seen, to the detailed consideration of the *Derniers Vers* as a single poem. The more obsessive the motifs within the poet's internal landscape, the more frequently they are bound to recur, striking the reader with the paradox of their coexistence and yet mutual incompatibility. The poetic text thus acquires characteristics that one normally associates with the visual arts, becoming a spatial complex rather than a temporal one.

Sharp dramatic contrasts are also set up in IV, again entitled 'Dimanches', and published in *La Vogue* as the companion poem to III. As in III, a 'real' scene outside the window (the girls going to church) sparks off the reverie about the loved girl and the possibility of marriage. The opposing tones of the poem—aspiration and irony—are more clearly separated in its earlier versions. The poem in the *La Vogue* version ended on an ironic disclaimer, similar to, if less amusing than, that on which III ends, and the manuscript of the poem sets out the theme of aspiration with a baldness that is not present in the subtler, but also more obscure, final poem:

> Oh, la misérable littérature
> Qui nous a raillé
> Qui nous a chansonné ces merveilles de la Nature!
>
> Oh, ce n'est pas souiller,
> Ce n'est pas une aventure,
> C'est entrevoir le Sacré
> De la pauvre, pauvre créature.

The narrator is not directly described, any more than he was in III. Instead, we have metaphors for him. When they are as varied as Hamlet, the polar bear of legend, and 'Le Grand Chancelier de l'Analyse', it is clear that we inhabit a fanciful world of the imagination.

Eliot rejected the Hamlet persona for his 'Prufrock' and replaced it by one ('am an attendant lord') more suited to the banality and realism of his character. Laforgue quotes Hamlet's words 'Alas poor Yorick!', and, since they are inseparable in any reader's mind from the scene of Ophelia's burial, the themes of death, regret, and remorse in the first stanza are echoed by this allusion. (The effectiveness of quotations in a foreign language is one of the devices to which Eliot may have been alerted in part by his reading of Laforgue.) Hamlet, clothed in mourning, is replaced by the startling whiteness of the polar bear, a comic figure suggesting the isolation, boorishness, and aspiration to purity of the poet. Bewilderingly, yet appropriately, since the three figures resist love while they aspire to it, the third figure is that of the philosopher, and tragedy, myth, and metaphysics inform this poem—as they do, in different ways, all of the *Derniers Vers*.

Not only is it impossible to visualize the narrator in a straightforward manner, but the narrative stance changes throughout the poem. Pronouns are mobile and imprecise. 'Je' moves to the impersonal 'on', and back again to the first person. 'Nous' at one point refers to the male sex, and at another to the narrator and the girl. 'Tu' refers to the miracle of love and to the girl. An unexplained pseudo-external voice is again present, this time encouraging the narrator to further confession: 'Voyons, confiez vos chagrins à votre vieil ami.'

The unexpected, unannounced transitions of the mind are also the subject of V and VI, on the face of it simpler poems whose titles encapsulate the theme of each—a heartfelt plea for equality in love in 'Pétition V', and a passionate attack on the cruelty of life in 'Simple

Agonie VI'.[1] But in detail the poems offer considerable interpretative difficulties. The narrator is more impersonal than he is in any of the other poems, and his object is still more shifting. In V he apostrophizes a number of disparate objects in quick succession: 'amour absolu', 'bouquets d'oranger', 'pauvre race', 'Circés', 'béatifiques Vénus', 'historiques esclaves'. In VI the opening line apostrophizes the 'paria', but it is not clear what this refers to: later it is linked with the world, but at the start of the poem the narrator seems a more likely attribution. The final stanza also introduces a surprising shift of viewpoint. The pseudo-external voice refers to the narrator in the third person: 'je' becomes 'un autre'. As in the last stanza of III, the 'new' narrator seems to represent the voice of common sense. He announces the death of the earlier speaker, addressing the reader in oratorical fashion, as if he were one of a crowd literally listening to his speech:

> Il prit froid l'autre automne,
>
> . . . .
>
> Il vint trop tôt, il est reparti sans scandale;
> O vous qui m'écoutez, rentrez chacun chez vous.

Despite the vivid dramatization of the 'voices' of the self in both these poems, the narrator is 'dematerialized'—nothing more, one might feel, than the split self talking aloud. Added to this, the context of the poem is extremely indefinite. One is reminded of Mallarmé in his late sonnets, but whereas Mallarmé blurred the outlines of the objects he described by placing them in a fading or unstable light, or by employing an impressionism that excluded all but a particular aspect, Laforgue offers movements, feelings, and values that exist almost in a void. His method is to rely on pronouns whose antecedents are hard to trace, and on nouns and adjectives that lack precise terms of reference: 'un monde plus mortel', 'les hommes de l'art', 'la chose qu'est la chose', 'l'instant le plus seul de la nature'. If pure introspection cannot 'create a character', no more can the rootless outpouring of emotion that we find in these two poems. The pleas of V, the denunciation of VI issue from a voice that is generalized, almost a twentieth-century 'universal singular'. Discontinuity is such a central and developed feature that one might say, as Todorov has said of Rimbaud, that disorder has become the

[1] The themes are stated plainly within each poem: 'Mon Dieu, que l'Idéal | La [woman] dépouillât de ce rôle d'ange! | Qu'elle adoptât l'homme comme égal!' (V); 'Et nous aurons beau la piétiner à l'envi, | Nous ne serons jamais plus cruels que la vie' (VI).

organizing principle, and that Laforgue has, like Rimbaud, bequeathed the discourse of schizophrenia as a model to twentieth-century poetry.[2]

Not that Laforgue is remotely like Rimbaud. His aim is to record the human, the earthbound, the psychological; he is no 'voleur de feu'. Nor do his poems trace the discontinuity of an ordinary stream of consciousness. The attempt was made in prose in 1887 by Laforgue's contemporary and acquaintance, Édouard Dujardin. His novel, *Les Lauriers sont coupés*, has the distinction of being both highly original and inescapably pedestrian.[3] It is quite clear, as we move from it to the *Derniers Vers*, that Laforgue offers us a heightened and structured version of the 'real', an intense and dramatized universe. Its language is 'de l'âme pour l'âme', as Rimbaud said the language of poetry should be.[4] Its tone, particularly in VI, can be almost manic, which reminds us of a Whitman in reverse, since Whitman's poetry was dedicated to the celebration of life; Laforgue, in this poem, to revolt and despair. Its obscurity has nothing in common with Kahn's in *Les Palais nomades*, where its function is to transport the reader to a climate of vague reverie. 'Simple Agonie' is obscure because it is in the nature of the consciousness to be so. This is poetry striking for its attempt to be faithful to the inner processes of the mind. No 'character' is created in the manner of Eliot's Prufrock: all such particularization is shunned. V and VI are experimental poems, hovering on the brink of the inexpressible, and representing merely one type of highly original interior monologue that Laforgue was prepared to try out. Their achievement is to give form to the diverse and discontinuous processes of the mind, and to do so in a language that reflects their instability.

[2] T. Todorov, 'A Complication of Text: The *Illuminations*', in T. Todorov (ed.), *French Literary Theory Today* (Cambridge, 1982).

[3] For an analysis of this interesting phenomenon, see Kathleen McKilligan, *Édouard Dujardin: 'Les Lauriers sont coupés' and the Interior Monologue* (Hull, 1977). It could be argued that the paradox is written into Dujardin's brief, which was to write 'le roman de quelques heures, d'une action banale, d'un personnage quelconque' (ibid. 90). The novel is remarkable in bringing together both external and internal impressions in a telegraphic style and in realistic confusion. It contains Flaubertian comedy at the spectacle of a naïve young man who repeatedly parts with considerable sums of money in the hope of sexual favours that continually elude him, or who is so little obsessed by the supposed object of his passion that he is diverted by every pretty woman who crosses his path. The sensibility of the hero, however, remains rudimentary, his sensations predictable and commonplace.

[4] A. Rimbaud, 'Lettre du voyant' to Paul Demeny, May 1871, *Œuvres complètes* (Paris, 1951), 255.

SYNTHESIS AND THE WAGNERIAN 'MOTIF' (*DERNIERS VERS* VII–VIII)

In 1931 Édouard Dujardin, by this time a keen admirer of Joyce's *Ulysses*, published an essay on the history of the interior monologue[5] in which he quite legitimately insisted on the originality of his own use of a stream-of-consciousness technique in his novel of 1887, *Les Lauriers sont coupés*.[6] He said that he had drawn the inspiration for the novel from the group of young poets who, in the 1880s, under the joint influence of Schopenhauer and Wagner, devoted themselves to registering the patterns of the inner life of the soul. He interpreted this as a natural reaction against preceding poetic movements, Romanticism and Parnassianism, but what was new, as he saw it, was that this inner life had its source in the unconscious. Reason and deduction were rejected: the poetic was felt to be inseparable from the musical, and had the capacity to enter all domains of literature—hence his 'poetic' novel. The music of Wagner was, Dujardin maintained, the gateway through which the members of this group could come to an understanding of Schopenhauer's ideas of the unconscious force dominating all life, and, for

[5] The essay was called 'Le Monologue intérieur, son apparition, ses origines, sa place dans l'œuvre de James Joyce et dans le roman contemporain'. The term 'stream of consciousness' was used by William James ('stream of thought, of consciousness or of subjective life') in *Principles of Psychology* (New York, 1902), 239. McKilligan (*Édouard Dujardin*, 90–7) casts doubt on the Wagnerian inspiration behind *Les Lauriers sont coupés* that is given such prominence in this essay, suggesting that the essay was written too long after the event to be reliable, and relying, herself, on the draft of a letter of 1888 that does not mention it. I see the situation somewhat differently. Behind the precise points made in the draft, it is not difficult to read a Wagnerian influence ('Toute l'action est dans l'âme du personnage'; 'la forme musicale de la poésie'). By the time Dujardin wrote his essay, Wagnerism had become a historical event that it was necessary to elucidate. Although he was personally disenchanted both with the aims and the achievement of Wagner by now, he felt bound to stress, perhaps with some nostalgia, the importance to him in his youth of an enthusiasm that had fired his whole creative personality. This interpretation does not at all invalidate the equal importance in the writing of *Les Lauriers* of purely literary influences and of Dujardin's own earlier writings.

[6] Joyce admitted the debt he owed to Dujardin, insisting that Dujardin's novel had given him the idea of the technique that he used in *Ulysses*, and writing the following inscription in a copy of the French translation of *Ulysses* that he sent to Dujardin: 'A Édouard Dujardin, annonciateur de la parole intérieure. Le larron impénitent, James Joyce.' (McKilligan, *Édouard Dujardin*, 20.) Dujardin, it appears, was not himself fully aware of the significance of the technique that he invented until Joyce imitated it, and, to do him justice, always recognized the aesthetic distance between his youthful experiment and Joyce's masterpiece. Presumably Mallarmé also did not guess the full importance that the form was to have, labelling it, as he did free verse, 'une trouvaille' (E. Dujardin, *'Les Lauriers sont coupés' suivi de 'Le Monologue intérieur'* (Rome, 1977), 198).

them, Symbolism meant freeing poetry of its intellectual yoke and placing it on the Schopenhauerian throne of music—to use Dujardin's image. He traced a direct analogy between musical 'motifs' and what he called the 'petites phrases directes du monologue intérieur'.[7]

Les Lauriers sont coupés was written, he said, in the ambitious hope of transposing the techniques of Wagner into literature: those musical motifs which expressed 'indéfiniment et successivement les "états" de la pensée, sentiment ou sensation'.[8] The life of the soul was conveyed by 'la succession indéfinie de courtes phrases donnant chacune un de ces états de la pensée, sans ordre logique, à l'état de bouffées montant des profondeurs de l'être, on dirait aujourd'hui de l'inconscient ou du subconscient . . .'[9]

It is well known that the combination of the poetic and the symphonic in Beethoven's Choral Symphony was a central inspiration to Wagner in his visualization of a totally new kind of opera, Germanic rather than Italian or French, which would have the power to evoke the character's often unconscious inner state. It would achieve this by what he called the symphonic treatment of a series of motifs that would be modulated, developed, and recapitulated in such a way as to give subtle expression to their interrelationship. Wagner's aim, to 'emotionalize the intellect', as he termed it, was a poetic as well as a musical concept. The interrelationship of art-forms was, of course, almost a commonplace by the time that Rimbaud, in his *lettre du voyant* to Paul Demeny (1871), wrote of poetry in musical terms: 'la symphonie fait son mouvement dans les profondeurs'. Laforgue, writing about the freedom of Rimbaud's poetic style, referred to both music and painting to illustrate his point: 'Avouons le petit bonheur de la rime, et les déviations occasionnées par les trouvailles, la symphonie imprévue vient escorter le *motif*; tout comme un peintre est amené là—à ce gris perle à propos de bottes, à ce géranium sans nécessité . . . tel le musicien avec ses harmonies qui ont l'air parasites.'[10] The fact that the 'symphonie', unanticipated because it springs from the depths of the personality and speaks to those depths, *supports* the 'motif' is precisely what Wagner was at pains to explain. Laforgue thus wrote *en connaissance de cause*, and would have met these ideas, no doubt filtered through Fourcaud, in the spring issues of the *Revue wagnérienne* of 1885. He understood that, however lacking in 'necessity' an element in Impressionist painting or in Wagnerian music might appear to be—and its apparently spon-

[7] Ibid. 258. [8] Ibid. [9] Ibid. [10] *MP* 130.

taneous randomness was essential to its effect—no such element was in fact either gratuitous or parasitic. Each related to the 'symphonie' that derived its being from 'les profondeurs'.

Both in Berlin and when accompanying the Empress Augusta on her tour of her regular watering-places, Laforgue had many opportunities to hear the works of Wagner. By the spring of 1886 they must have been very familiar to him, the most frequently heard being, apparently, *Siegfried* and *Die Walküre*. 'Entendu dix fois la Walkürie, Siegfried',[11] he said, referring to a matter of weeks. When he met Dujardin in March, and spent most of three days and nights, Dujardin reported, talking about their common poetic interests, he gave Dujardin the impression, shared by Kahn, that he was more interested in psychology than in musicality: 'tandis que je parlais expression musicale, Laforgue répondait expression psychologique.'[12] One can well imagine discussions in which Laforgue tried to draw his new friends Dujardin and Téodor de Wyzewa, a regular contributor to the *Revue wagnérienne* that Dujardin edited, towards a recognition of the importance of the psychological to balance and give meaning to the musical. Dujardin and Wyzewa probably, rather like Kahn, undervalued the 'substance' that was to be treated musically—witness the charming but banal sonnets that they contributed to the *hommage* to Wagner which end the sequence begun by Mallarmé and Verlaine. Kahn's and Dujardin's statements have done Laforgue a disservice, in that they have drawn attention away from the musicality of his free verse. Laforgue insisted on various occasions on the importance that he attached to musicality in poetry, one of the most notable being his criticism of Corbière for his lack of it, contained in a letter of 1885 to Leo Trézenik: 'Corbière ne s'occupe ni de la strophe ni des rimes (sauf comme un tremplin à concetti) et jamais de rythmes, et je m'en suis préoccupé au point d'en apporter de nouvelles et de nouveaux; j'ai voulu faire de la symphonie et de la mélodie, et Corbière joue de l'éternel crincrin que vous savez.'[13]

The 'psychological' was extremely important to Wagner, as Fourcaud explained in his survey of Wagner's aesthetic theory, which Laforgue probably knew: 'Ainsi l'orchestre devient une source de vie scénique et l'agent par excellence de la *psychologie du drame*;' and again: 'l'absolue conformité de la musique aux paroles et la liberté de l'orchestration, chargée de fournir à l'action une atmosphère, de dégager le fluide lyrique et *d'éclairer le dessous des caractères*, voilà les deux principes de la

[11] *LA* 166. [12] E. Dujardin, *Les Premiers Poètes du vers libre* (Paris, 1922), 59.
[13] *OC*, v. 137.

méthode wagnériste.'[14] The psychological aim in no way excluded the musical for Wagner; in fact, they were complementary.

If we turn to the seventh and eighth of the *Derniers Vers*, both of which we have already considered for their transposition of elements drawn from the *Fleurs*, we can easily see the importance of something very like the recurrent Wagnerian motif. However spontaneously the motifs appear to rise from the text, we can be sure that Laforgue saw them as the structural devices that they clearly are. In 'Solo de lune VII' musicality is already suggested by the title, which implies a singer or a solo instrument. In Chapter 5 we noted three personae: the aesthete of the night journey, the lover, and the ironist. The motif of the ironist pervades the poem in brief, playful intrusions, and, indeed, both opens and closes it ('Ma carcasse est cahotée'; 'J'eusse été le modèle des époux! | Comme le frou-frou de ta robe est le modèle des frou-frou')—a motif of a kind not typical of Wagner, but frequently present in his more whimsical successor, Richard Strauss. The aesthete's receptivity to, and enjoyment of, the subtly changing beauty of the moon- and star-lit night scene provides another motif that is modulated throughout the poem. The figure of the lover is the most complex and shifting. He has three essential motifs: that of nostalgia for the past; a counter-motif of a more carefree and fantastic character, addressing itself to future imaginings; and against both of these is set a note of urgent and immediate anxiety.

Each motif develops through the poem—that is, it follows a modulated sequence. While the motifs are contrasted with each other, they also interrelate, introducing or provoking one another. It is the fact that the cotton- and saw-mills of the urban landscape are left behind that causes the narrator's monologue also to take flight from the mundane. It is the beauty of the moon- and star-lit night that reminds him of his aspiration to the absolute in a different sphere from the aesthetic—that of human love—and which therefore leads directly to the narration of a past unfulfilled love. It is the dark Wagnerian forests that invite fantasies of a primeval and fulfilling love; it is the chill night air that causes fear for the girl's safety, and gives rise to premonitions of death. The journey, an active element in the monologue as well as a background to it, leads the meditation to the central areas that must be given emotional expression. If the poem ends on a delightfully laconic disclaimer, the reader, who has been offered a variety of responses to the situation, is

[14] Louis de Fourcaud, in the *Revue wagnérienne*, Mar. 1885 (my emphases).

not likely to forget the earlier insistent 'comprenez-vous?', which has been addressed directly to him; nor the dramatized glimpses of pathos or beauty, aspiration or loss. In the free association of the narrator's meditation, we are given a configuration of tones and textures, as well as a synthesis of past, present, and future moments. Repeated phrases ('La lune se lève | O route en grand rêve', or 'Soigne-toi, je t'en conjure') act as refrains around which the symphonic structure takes shape.

The regular and irregular in the verse-form work in conjunction with each other—each gaining its resonance to some extent from the presence of the other. The 'quatrain populaire' provides the expected regularity against which the free verse is set. It is used by the narrator for recapitulation. (The first such stanza is placed directly after the authorial injunction: 'récapitulons!') The well-rehearsed demands metrical regularity, but often deserves also the sardonic sting in the tail that Laforgue does not hesitate to provide:

> Nous nous aimions comme deux fous,
> On s'est quitté sans en parler,
> Un spleen me tenait exilé,
> Et ce spleen me venait de tout. Bon.

Both the octosyllabic quatrain and the *rime embrassée* rhyme scheme have been destabilized by the ironic addition of 'Bon'. The flux of the present is rendered in the fluid verse that one finds—in the final three stanzas, for example—matched by tentative expressions of supposition ('Oh! si à la même heure, | Elle va de même le long des forêts') and of artistic self-questioning ('O Solo de lune, | Vous défiez ma plume'). So subtle are the modulations in the text that Clive Scott is totally convincing when he finds them even in the different tones of exclamatory interjections: the staccato, unemphatic 'O', the nostalgic, past-directed 'Ah!', and the anticipatory 'Oh!'.[15]

The sense of integration achieved in the texture of the writing can be illustrated by looking at the opening stanza of the poem.

> Je fume, étalé face au ciel,
> Sur l'impériale de la diligence,
> Ma carcasse est cahotée, mon âme danse
> Comme un Ariel;
> Sans miel, sans fiel, ma belle âme danse,
> Ô routes, coteaux, ô fumées, ô vallons,
> Ma belle âme, ah! récapitulons.

[15] C. Scott, *A Question of Syllables* (Cambridge, 1986), 166–75.

The self and the environment are immediately set in apposition. In the first line the subjective 'Je fume' is clearly less significant than the environmental 'ciel'. The next contrast narrows the focus to the two parts of the self, physical and spiritual. The physical 'ma carcasse est cahotée' is subordinated to the spiritual 'mon âme danse'. In both cases, therefore, the spiritual dominates, radiating from the environment (*ciel*) to the poet. The activity of the soul contrasts with the inactivity of the body, showing where the emphasis of the poem must finally lie. The word 'âme', the climax of these three lines, and the central focus of the entire stanza, derives meaning from the whole context, not merely from its evident contrast with 'carcasse'. It is not only spirit as opposed to body, but a spirit of a peculiar 'je m'en foutiste' variety, and its opposite is not simply a Baudelairean *charogne in potentia*, but a body borne on a royal carriage and surrounded by the glories of the universe. It is in the context of these interrelationships that we attribute meaning to what follows: to the Shakespearian Ariel reference, to the repeated 'belle' that half ironically qualifies the soul. And, with this patterning in mind, we move on to the simple regular rhythm of the little tale of unfulfilled love that follows.

Narrative taken from the *Fleurs* plays an important part in 'Solo de lune' and in the next poem, 'Légende'. Both poems are concerned with an unfulfilled relationship, but whereas 'Solo de lune' is about an unspoken 'ideal' love ('Nous nous aimions comme deux fous'), 'Légende' presents a one-sided situation. The girl has been jilted, and rejects the poet's love because 'Avec vous autres rien ne dure', and because 'On n'aime qu'une fois'. These *idées reçues* that she quite realistically offers the narrator represent the two opposing motifs of the poem: the 'real' and the 'ideal'; but, as in the body–soul opposition in 'Solo de lune', we are not left with these bare contrasting concepts. Ambiguity in the attitudes of the narrator and the girl are present from the outset. The narrator stands for the ideal, but at the start of the poem he only half believes in it. Apathy is his heraldic device. He listens disconsolately to the girl's tale of seduction and betrayal, suffering from the banality of which she is quite unaware. Her narration trails away into the clipped, staccato, ready-made phrases quoted by the narrator as he turns them over in his memory: 'Avec vous autres rien ne dure', which is the natural conclusion of her experience, and the equally trite 'On n'aime qu'une fois', which is her excuse for rejecting the narrator. Investing the cliché with its original meaning, he is transported into an imaginary golden world of fidelity, gentleness, and value, a world of true love, peopled by heroines of legend, of which he now feels himself to be a part. Thus the girl, whose limited understanding is painfully 'real', has been the catalyst for the poet's dream of the 'ideal'. Ironic in every detail, the poem explores and illuminates an inner landscape while it balances with great delicacy the pitiful reality and the romantic dream. As in 'Solo de lune', Dujardin's 'petites phrases directes du monologue intérieur' have provided the motifs which, when orchestrated, create a complex whole, so that, in Laforgue's words, 'la symphonie imprévue vient escorter le motif'.

Discontinuity is thus countered in the *Derniers Vers* by a desire for synthesis, inspired in part by Wagner's operatic aims and method—a desire that Laforgue also found expressed in Impressionist painting, of which he wrote: 'Le tout est une symphonie qui est la vie vivante et variante, comme "les voix de la forêt" des thèmes de Wagner en concurrence vitale pour la grande voix de la forêt.'[16] This desire had a place even in the fractured discourse of a poem such as 'Simple Agonie', in which Laforgue referred to his work as a melody, characterized it by its unity, and expressed the wish that it might capture the true nature of existence.[17]

[16] *MP* 137.

[17] 'Oh! que | Devinant l'instant le plus seul de la nature, | Ma mélodie, toute et unique, monte, | Dans le soir et redouble, et fasse tout ce qu'elle peut | Et dise la chose qu'est la chose.'

10

The *Derniers Vers*: 'L'instant pris à la gorge'

THE ASSAULT ON THE READER (*DERNIERS VERS* IX-XI)

BORGES pointed out that, as the taste of the apple lies in the contact of the fruit with the palate, and not in the fruit itself, so poetry lies in the meeting of poem and reader.[1] Laforgue, who was so receptive to the innovations in Baudelaire's writing,[2] who admired his most unusual images 'ces comparaisons crues . . . palpables . . . en un mot américaines'[3]), who aimed himself to 'faire de l'original à tout prix',[4] must have been very conscious that all innovation launches an assault on the reader. When he gauged the subtlety of the exclamatory 'Hélas!' in Baudelaire's line: 'Veuve d'Hector, Hélas! et femme d'Hélénus' ('Cet Hélas! n'est ni poncif racinien ni une cheville mais d'une subtilité touchante et grande'[5]), and when he praised his unconventional use of the word 'très',[6] and used it himself in more unorthodox ways still in phrases such as: 'C'était un très-au vent d'octobre paysage'[7]—he clearly knew that even the innovatory use or placing of a single word could constitute such an assault. No poet of Mallarmé's generation could be unaware that poetic creation was self-creation, though it was Yeats who perhaps put the point most amusingly: 'The friends that have it I do wrong | When ever I remake a song, | Should know what issue is at stake: | It is myself that I remake.'[8]

Stylistic innovation is a new way of seeing, which imposes its novelty of vision on the reader. The most striking examples of such assaults in the *Derniers Vers* are perhaps those opening stanzas, of which X is a good example, where minimalist sense-units remove all sense of continuity:

[1] J. L. Borges, Preface to *Obra Poetica* (Buenos Aires, 1969).

[2] This area is discussed above, in 'The Literary Situation'.

[3] J. A. Hiddleston, *Essai sur Laforgue et les 'Derniers Vers' suivi de Laforgue et Baudelaire* (Lexington, Ky., 1980), 97.

[4] *AH* 821.

[5] Hiddleston, *Essai sur Laforgue*, 108.

[6] Ibid. 111.

[7] 'Complainte d'un autre dimanche'.

[8] W. B. Yeats (1907), *Collected Works*, i (New York, 1989), 551.

O géraniums diaphanes, guerroyeurs sortilèges,
Sacrilèges monomanes!
Emballages, dévergondages, douches! Ô pressoirs
Des vendanges des grands soirs!
Layettes aux abois,
Thyrses au fond des bois!
Transfusions, représailles,
Relevailles, compresses et l'éternelle potion,
Angelus! n'en pouvoir plus
De débâcles nuptiales! de débâcles nuptiales!...

Until the final negative infinitive of defeat, the stanza contains no verb. Rimbaud must surely have alerted Laforgue to the dramatic effectiveness of the repeated apostrophe. The extreme emotion that is registered here in a series of urgent and explosive exclamations forces the reader to react: to share enthusiasm, apprehension, disgust; to sympathize, to disbelieve, to mock. Mounting tension is conveyed by the accumulation of paratactic epithets: two sets of ternaries—the first accompanied by adjectives, the second the barest of ejaculations. The passage has meaning only if the reader relates each unit to an absent narrator. Each accompanying adjective in the first group sets up a contradiction: the richly coloured geraniums are diaphanous (suggesting the paradox involved in feminine attraction); the warriors are not bold but wily, obtaining their victories by evil spells; the 'sacrilèges monomanes' are sacrilegious, presumably because the object of their obsession is not religion but sex. The three nouns that follow them are similarly disorientating, yoked together by the exclamation mark that follows the last: the first two offer a semantic antithesis but a phonetic parallel in 'emballages, dévergondages', while 'douches', moving to the concrete after the two earlier abstract nouns, leads on to the sordid sequel to the sexual 'dévergondages'. This whole series of juxtaposed epithets is no 'bibelot d'inanité sonore':[9] it prepares the way for the metaphor of the wine harvest for orgiastic love-making: 'Ô pressoirs | Des vendanges des grands soirs!'—a metaphor that is given intensity by the 'Ô' that parallels the opening apostrophe, and fluidity by the first *enjambement* in the poem. Each element retains its separateness, however, which matches on a thematic level the depiction of sexual love as an illusory and temporary union between discrete entities. The pattern of the first half of the stanza is repeated in the second, which again consists of a list

[9] Stéphane Mallarmé, 'Ses purs ongles', sometimes known as the sonnet 'en yx'.

of unexplained epithets, all with sexual connotations, and all, in Baudelaire's phrase, 'laissant carrière à la conjecture', whether it be the Bacchic thyrsus, the churching ceremony, or the infants' layettes. Thus we move to the climax of the stanza, the inevitable result of the sexual adventure, the emphatically repeated 'débâcles nuptiales'.

Dramatic openings that, like this stanza, launch a simultaneous attack on the emotive and interpretative levels are not uncommon in Laforgue. There is the 'blocus sentimental' flung at the reader in the opening line of 'L'Hiver qui vient', which is also a pun on 'blocus continental'; the evocation of autumn as a 'chute des feuilles, des Antigones, des Philomèles' in IV; the definition of absolute love as a 'carrefour sans fontaine' in V; the address to the unspecified pariah—earth or poet?—of the first line of VI; or the enigmatic opening of VIII, where 'Armorial d'anémie' symbolizes the poet's Hamletian indecision.

Such epigrammatic exclamations, which are often paradoxical and unnerving, are valuable because they offer fertile collisions of meaning from which the poem can develop. Sometimes they are repetitions of formulas first discovered for the *Fleurs*, as in III and IV. They are valuable also because they act as starting-points from which the poem can diverge sharply. In X, for example, after the violent assault of the stanza that we have examined, Laforgue moves to an intimate address to the reader, sharing with him his still contradictory and inconclusive dreams. The tone is no longer bombastic and shocking; it is a tête-à-tête of the kind he admired in Verlaine, but one which he, unlike Verlaine, intellectualizes. There is a natural move to the first person as the reader becomes the explicit confidant. The bombastic opening has made us receptive and even vulnerable to this different kind of assault, which appeals to our idealism and sensibility. We have first the dream of innocent domesticity ('O ma petite mienne, ô ma quotidienne'), which is followed by the evocation of a love whose romantic status is maintained by mystery and absence ('Oh, qu'elle est loin! Oh, qu'elle est belle! | Oh! qui est-elle? A qui est-elle?'). Both, while appearing opposites, describe the poet's insubstantial reveries. Only 'trains manqués' have this picturesque quality. They are not unrealizable but unrealized dreams, which inspire nostalgia and anticipation because they are both safely in the past[10] and capable of resuscitation.

[10] The evocation of the past held a privileged position in Laforgue's aesthetic, as the following note shows: 'Il faut dire qu'une description porte la note de votre cœur,—et le moment où vous avez votre cœur, c'est non pas devant la chose crue, encombrante, mais quand plus tard, songeant, seul, nostalgique vous évoquez l'éphémère.' (*FV* 267.) For a

Oh! qu'ils sont pittoresques les trains manqués!...

Oh! qu'ils sont 'A bientôt! à bientôt!'
Les bateaux
Du bout de la jetée!...

De la jetée bien charpentée
Contre la mer,
Comme ma chair
Contre l'amour.

Such intimate glimpses of the narrator's aspirations and reservations form a central element in most of the *Derniers Vers*. They are the fragile nucleus that is protected by the surrounding rhetoric, declamation, parody, all of which are equally elements of the poet's response to life. IX and XI are unusual in taking us directly to this inner level:

Oh! qu'une, d'Elle-même, un beau soir sût venir
Ne voyant plus que boire à mes lèvres, ou mourir!...
(IX)

The simplicity of such an opening is, however, more apparent than real. If the dream is offered first, it is clearly in danger of demolition. Laforgue's third, and most obvious, method of assault on the reader is, of course, the by now thoroughly familiar one of irony, which offers the reader an unstable and elusive narrator. In the case of Laforgue, it is not a question of a narrative technique used to avoid over-simplification or omniscience. It is a method that assists self-expression, and the instability, surrounded by a multitude of very real sensations, is precisely the situation that he wishes to convey. As a result, the poems are explorations that cannot reach their goal.

Derniers Vers IX is a wish-fulfilment fantasy, similar to the Pierrots' speeches—if more expansive, and figuring a more obviously vulnerable narrator. XI is a denunciation of the irremediable human situation, dominated by the 'hasard' that no throw of the dice can abolish. IX and XI are neither 'sous-conversations' (contrasting with a real and very different conversation) nor are they 'préconversations' (capturing sensations and emotions before they are consciously formulated): they are sustained interior monologues, whose purpose is precisely to replace a conversation. Highly dramatized, they lead in each case to a triumphant and, at the same time, ridiculous finale. Both poems are about the desire

discussion of Laforgue's work from the angle of his treatment of time, see *Hidd. LA* 51–65.

for the absolute in love. In IX the narrator imagines himself as the recipient of an idolatrous love; in XI he imagines himself rejecting such a love merely because he owes it to chance: 'Si elle avait rencontré seulement | A, B, C, ou D, au lieu de Moi, | Elle les eût aimés uniquement.' The poems are mirror images of each other. In the first we are given a pathetically destitute narrator; in the second an arrogant anti-self. That both these ridiculed stances stem from experience can be seen from their reflection in such autobiographically based characters of the *Moralités légendaires* as Lohengrin or Hamlet, as well as from Laforgue's personal writings.[11]

How deeply the Schopenhauerian and Hartmannian vision of chance as the driving force of the universe coloured his perception of life can be seen in a letter, written at the time of his engagement to Leah, which might otherwise appear strange: 'Oui, tout est hasard, car . . . n'y eût-il pas eu de Leah, il y aurait eu une Nini, et ainsi de suite. C'est pourquoi il nous est enjoint de nous attacher à la première que le hasard nous présente, et nous l'aimerons seule, car c'est la première et nous ne rêverons pas à une autre.'[12] This is the subject-matter of XI, expressed here in all seriousness, and from the angle of acceptance of the relationship. In XI it is satirized, and Laforgue makes quite sure that the narrator will be unacceptable to the reader in a final stanza of crude brashness:

Et, si je t'avais aimée ainsi,
Tu l'aurais trouvée trop bien bonne! Merci!

The last stanza of IX has a similar function—and, of course, irony has been present throughout in the shrill simplicity of the girl's language, which represents the extremity of the sentiment that the poet wishes to inspire. The last stanza of IX is taken from *Fleurs* II, where it ran like this:

—Ainsi, elle viendrait, évadée, demi-morte,
Se rouler sur le paillasson qu'est à ma porte!

In IX Laforgue changes it to increase the parody: 'le paillasson que j'ai mis à cet effet devant ma porte'. In each of these poems, which illustrate

[11] These develop in prose many of the stances that we have observed in Laforgue's verse. They will appear in full in volume 3 of the 'L'Âge d'homme' edition of Laforgue's works, edited by D. Grojnowski and J. Debauve.

[12] *OC*, v. 163. This insistence on the role of chance does not invalidate emotion in the least. The letter to Laforgue's intimate friend Théo Ysaÿe ends: 'Je la verrai dans une demi-heure. Cette minute me fait palpiter le cœur, et dans quarante ans je penserai combien longue à venir fut cette minute.' (Ibid. 164.)

the ironic method at its most dramatic, the narrator is mocked by the sheer exaggeration of either his desires or his demands, and yet he has moved too close to the workings of the inner mind for the reader's comfort. Jean Genet, whose dramas express ritualized fantasies of the self, described what he saw as the essence of the theatrical experience: 'Je vais au théâtre', he wrote, 'afin de me voir, sur la scène (restitué en un seul personnage ou à l'aide d'un personnage multiple et sous forme de conte) tel que je ne saurais—ou n'oserais—me voir ou me rêver, et tel pourtant que je me sais être.'[13] Laforgue's theatrical poems also offer a mirror that does not permit evasion.

A further assault on the reader, and one which is inseparable from irony, results from Laforgue's 'quotational' style. The narrator in the *Derniers Vers* visibly confronts the difficulty of attempting to express personal feelings in a language created by others, its formulas constantly repeated, a language that he cannot 'possess' but merely 'use'. Trite *idées reçues* exemplify this state of affairs, seeming to destroy any potential that language might have for individual self-expression. This is not the whole story, however. In Laforgue's poetry the permutations of these formulas are so complex, the ironies so 'unstable'[14] and, therefore, rich and evocative, that in the *Derniers Vers* even the cliché can be rejuvenated, as we saw in 'Légende VIII'. We translate it back, as it were, to what might have been the original desires, aspirations, or fears that a crude and blunt usage has distorted, and we see in the 'usure du langage'[15] the 'usure' also of human life. To force the reader on such a journey is to use language with considerable authorial refinement. It involves drawing attention to the wider areas of tone and gesture, which Laforgue signals by devices such as italics, suspension points, exclamation marks, and, of course, as we shall see in the next section, by his choice and use of metre.

Francis Scarfe drew attention to Laforgue's originality as regards genre by claiming that he 'invented a new kind of dramatic monologue usually known as the interior or internal monologue, close to common speech'.[16] This seems an accurate description of the situation. Laforgue's monologues are inherently dramatic, depicting conflict both

[13] J. Genet, 'Comment jouer les bonnes', *Les Bonnes* (Décines, 1963), 10.

[14] Wayne C. Booth, in *A Rhetoric of Irony* (Chicago, 1974), divides the concept of irony into the two categories 'stable' and 'unstable'.

[15] The phrase is applied to Laforgue by John E. Jackson, *La Question du moi* (Neuchâtel, 1978), 58.

[16] F. Scarfe, 'Eliot and Nineteenth-Century French Poetry', in G. Martin (ed.), *Eliot in Perspective* (London, 1970), 53.

within the self and between the self and the external world, which, by its neglect of the narrator, its cruelty, or its inadequacy, has promoted the divisions in his mind that are the subject of the monologue. They are also clearly directed towards the inner world of the consciousness, and aim to bring to light the play of the imagination at work, the private 'tête-à-tête avec soi'. Introspective and self-questioning, as we have seen, they uncover areas that we normally connect with the psychic personality: the depressive, the manic, or the obsessional. They derive, as Grojnowski points out, from 'un projet d'exploration', and expose 'la conscience au lieu du surgissement'.[17] Valéry Larbaud, in his Preface to *Les Lauriers sont coupés*, defined the interior monologue as 'la pensée intime en formation', and this is precisely the sensation that certain of the *Derniers Vers* convey. Prosody itself is used, as Grojnowski rightly insists, as an 'instrument d'introspection'.[18] Destructive as well as exploratory, some of the *Derniers Vers* show syntax dissolving, and rhetoric for what it is. They undermine poses and subvert logic. While being less naturalistic than Dujardin's prose in his novel *Les Lauriers sont coupés*, they convey a more vivid sense of dissolution, and offer a more fundamental assault on language itself, not least because they are written in verse rather than prose. Kahn, it seems, had not fully realized this fact when, in August 1886, he published *Derniers Vers* I and II in italics, as though they were prose poems. Laforgue had to point out the difference to him and pass it off as a joke: 'Je ne te signalerai pas deux coquilles, mais me plaindrai que tu m'as imprimé en italiques. *C'étaient des vers, Monsieur*!'[19] The fact that Kahn did not really see what Laforgue's enterprise actually was, even after he had had it explained to him, emphasizes the contemporary strangeness of the new form, and casts further doubt on Kahn's claim to have invented it.

In an interesting recent study of the dramatic monologue in Laforgue, Mallarmé, and Valéry[20] Elizabeth Howe takes issue with the statement by Francis Scarfe that I have quoted, and separates the interior monologue from the dramatic monologue, which is the genre in which she places Laforgue's poems. Basing herself on Dujardin's definition of interior monologue, she naturally sees Laforgue's poetry as doing something different. The technique that Dujardin is describing, since he is chiefly concerned with Joyce's *Ulysses* while making claims for his own

[17] D. Grojnowski, *Jules Laforgue et l'originalité* (Neuchâtel, 1988), 173.

[18] Ibid. 174.

[19] *LA* 215 (my emphasis).

[20] Elizabeth Howe, *Stages of Self: The Dramatic Monologues of Laforgue, Mallarmé and Valéry* (Athens, Oh., 1990).

novel, is now commonly described as 'stream of consciousness' or 'automatisme verbal', a technique that was not properly developed until well after Laforgue's day. Problems are more apparent than solutions in the attempt to categorize Laforgue's work, partly because, as Howe says, Laforgue is a 'transitional figure',[21] and partly because his work is, at every moment of his brief poetic career, so multifarious. We are concerned here with the *Derniers Vers*, and even within this small group of poems we have problems if we attempt a more precise definition than Scarfe's of the genre to which the twelve poems belong. We have first to admit that II is not a monologue at all. I is perhaps best described as a tone poem, evoking a seasonal moment that is also an 'état d'âme'. The degree of obscurity of the remaining ten poems is very variable: V and VI are extremely obscure and come close to a stream-of-consciousness technique; whilst VII is scarcely ever obscure, and, although it undoubtedly traces a complex 'tête-à-tête avec soi', it is clear that this is a stylized and selective variety of the form. Narrators vary between being acceptable personae for the poet (VII and XII), being little more than a bewildering series of masks (IV), and being scarcely defined at all (V and VI). The outside world can form a central and relatively stable background to the whole poem (I and XII). It can be a changing agent in the monologue (VII), a liminal image from which the poem develops (IV), a central image around which it converges (III), a single reference in the final stanza (IX), or no more than a vision in the mind's eye (XI). Similarly, a story-line may be offered (VII, XII), denied (I, V), or follow a number of impressionistic and fanciful paths (III, VI, IX, XI).

These differences are matched by the paradoxes that face us if we consider either the monologist, his narrative, or his language. The mask or persona that Laforgue adopts appears a retreat from the self, but it is also, and more importantly, an indirect approach to that self.[22] The ironic tone of the narrative appears negative and critical, but it is also, as we have seen, 'a means to indirect statement which increases rather than "undercuts" the felt pathos'.[23] The concentration on a specific theme

21 Ibid. 183.

22 My view differs from Howe's interpretation of Laforgue (ibid. 55–91) on two significant points: the possibility of reader identification with Laforgue's narrators, and the sense of interiority conveyed by Laforgue's poetry. Howe writes: 'Self-conscious and self-critical in the extreme, *Laforgue's personae make it impossible for the reader to identify with them*' (ibid. 75); and 'this view of language as repetition *undermines the very concept of interiority itself*' (ibid. 81) (my emphases).

23 P. Brooks, 'The Rest is Silence: Hamlet as Decadent', in W. Ramsey (ed.), *Jules Laforgue: Essays on a Poet's Life and Work* (Carbondale and Edwardsville, Ill., 1969), 108. Interestingly, M. Bakhtin (*Esthétique et théorie du roman* (Paris, 1978), 101) made

may seem contrived,[24] but it actually mirrors rather accurately the patterns of the mind in states of intense emotion. Free verse appears stylized, but, like the Freudian slip, its patterns of association may well have more to do with the inner 'modulations individuelles'[25] than does the language of prose.

What is clear is that the monologues of the *Derniers Vers* are intent on exploration of an inner area that cannot be fully known, but to which approaches can be made. Since these approaches reveal such a range of bewildering impulses and refractions, the reader may recognize only with some sense of shock the 'ESPACE de possibilités'[26] that has been thrust before his eyes. Laforgue registered the situation with amused resignation, and incorporated this irony into the fabric of his writing—witness the pun on 'peine' and the contrived active verbs that are used to justify inaction in III:

> Bref, j'allais me donner d'un 'Je vous aime'
> Quand je m'avisai non sans peine
> Que d'abord je ne me possédais pas bien moi-même.

But this was only one side of the picture, as the following statement, which unconsciously echoes Mallarmé, shows: 'Avant je ne m'intéressais qu'aux choses elles-mêmes—maintenant je m'intéresse surhumainement à l'ombre que font les choses;'[27] or this description of his hero in a fragment from one of his projected novels: 'Il passait des journées à s'épier en dedans, avec l'immobilité des moines du Mt Athos.'[28]

THE USES OF FREE VERSE (*DERNIERS VERS* XII)

It was to be expected that the free verse of the *Derniers Vers* should convey a strong sense of the line as unit, since this was Laforgue's justification for writing in free verse rather than in a poetic prose. In the first stanza of XII, the poem that we shall take as an example of Laforgue's

Laforgue one of the few exceptions to his conclusion that ironic distancing and the language of poetry were incompatible. The others were Horace, Villon, Heine, and I. Annenski.

[24] Howe considers that Laforgue's monologues are 'dramatic monologues' rather than interior monologues, because they are 'concentrated on a specific problem', 'aimed at an effect' (Howe, *Stages of Self*, 66).

[25] From Mallarmé's definition of free verse. See Ch. 6 n. 13.

[26] P. Valéry, *Cahiers*, ii (Paris, 1974), 277.

[27] *FV* 276. Written late in 1885.

[28] Ibid. 104.

latest free verse, every line, however subdivided, is a rhythmical unit. The pause at the line ending is emphasized by punctuation in six of its eight lines, and in the remaining two we do not experience the dislocation of the true *enjambement*. The phrasing has remained intact. Each line delivers a separate message before it dissolves into the reader's apprehension of the larger whole of the poem. One is reminded of the fact that 1887 was the year, and Paris the place, in which the first moving pictures were developed by Augustin Le Prince. The sequence of stills that merge to provide a moving image of reality offers an attractive parallel to the free-verse process:

Noire bise, averse glapissante,
Et fleuve noir, et maisons closes,
Et quartiers sinistres comme des Morgues,
Et l'Attardé qui à la remorque traîne
Toute la misère du cœur et des choses,
Et la souillure des innocentes qui traînent,
Et crie à l'averse: 'Oh! arrose, arrose
'Mon cœur si brûlant, ma chair si intéressante!'

'J'oublie le nombre des syllabes'[29] was, as we have seen, Laforgue's euphoric comment on his new poetic method, but quite what he meant by this celebration of amnesia has puzzled critics—and understandably so, since his lines so often fall significantly short of expected metres such as the alexandrine, or extend beyond them, with apparently calculated humorous effects, which suggest not only Laforgue's own self-consciousness concerning metrical expectations, but also his anticipation of these in his readers. There is no real problem, however. Laforgue can both enjoy his new-found freedom, which allows him greater accuracy in the translation of his affective inner world, and exploit deviations from the poetic norm as sources of further psychological subtlety. Similarly, he can make use of rhyme and near rhyme for a variety of purposes, from musicality to provocation. In the stanza quoted above, rhyme emphasizes line endings in the ternary rhyme 'close', 'chose', 'arrose', but we are also surprised by its absence ('Morgues' has no rhyming partner but is echoed internally in 'remorque') and by a rhyme that links lines as far apart as the first and the last.

Clive Scott compares Laforgue's 'free verse of the line'[30] to Régnier's, and contrasts it with the work of Apollinaire, but Apollinaire's poems that stress dislocation, such as 'Zone' or 'Le Voyageur', use a strongly

[29] *LA* 193.

[30] C. Scott, *Vers libre* (Oxford, 1990), 214.

linear free verse, rather similar to Laforgue's. Both poets also extend the sense of the line as unit by their use of the one-line stanza. After the first stanza of XII, Laforgue gives us an eleven-syllable 'stanza', extended by suspension points that merely remind the reader of the disharmony of the imparisyllabic line:

> Oh, elle, mon cœur et ma chair, que fait-elle? . . .

The question is left hanging, unanswerable. This effect is created in part by the line's isolation, and in part by its irregularity. One has only to provide the line with an extra—invented—syllable to give it back the rhetorical status of the alexandrine:

> Oh, Angèle, mon cœur et ma chair, que fait-elle?

The line might now conclude a stanza by Hugo.

The isolated line is used by Laforgue for the sudden, urgent appeal, whether it be this forlorn enquiry, the plea of the following regular, but deliberately unharmonious, alexandrine (3, 3, 3, 3):

> Soigne-toi, soigne-toil! pauvre cœur aux abois.

or the desperate brevity of the later isolated line:

> Oh, que nous mourrons!

Couplets and quatrains have a quite different function. In each of the two couplets in this poem we move from a summary of the background situation to the reiteration of an emotional stalemate:

> (Langueurs, débilité, palpitations, larmes,
> Oh, cette misère de vouloir être notre femme!)
>
> Nuit noire, maisons closes, grand vent,
> Oh, dans un couvent, dans un couvent!

Rhyme or near-rhyme stresses the isolation of the couplet, while in the first example brackets highlight subordination. The quatrain is used also for recapitulation or, in other poems, for simple narrative. The 'quatrain populaire'[31] was the only regular element that Laforgue intended to retain in his free verse, and in many of the *Derniers Vers* he quite simply transposes quatrains from the *Fleurs*. In poems such as XII, which contain no elements from the *Fleurs*, he clearly creates groups of four lines—sometimes set out separately from the rest of the poem, sometimes not—which have the simplicity and near regularity of the

[31] *LA* 194.

'quatrain populaire'. Take the following example, which is not set out separately, but which conveys a sense of unity by its description of past calm in lines of close, though not identical, syllabic lengths (9, 9, 10, 8):

> Un couvent dans ma ville natale
> Douce de vingt-mille âmes à peine,
> Entre le lycée et la préfecture
> Et vis-à-vis la cathédrale . . .

The regular, song-like quality of these lines will later be broken by wildly divergent line-lengths, which will oppose the poet's present emotional turmoil to this atmosphere of past stillness. The 'déjà connu' demands regularity. It can, of course, be a less acceptable familiarity than this nostalgic glimpse of a past that, ironically, Laforgue never knew.[32] It may have negative overtones of the obsessional and the insoluble, as in the following four lines, which are set out as a stanza:

> Oh! je ne puis voir ta petite scène fatale à vif,
> Et ton pauvre air dans ce huis-clos,
> Et tes tristes petits gestes instinctifs,
> Et peut-être incapable de sanglots!

Rhyme and liminal phonetic repetition stress regularity, while simple accumulation gives the quatrain a childlike quality, which contrasts with the complex irregular movement of the main part of the poem.

The poem begins, as we have seen, with the evocation of a bleak external world that is also an internal world. As in X, this impression is first created by an abundance of nouns:

> Noire bise, averse glapissante,
> Et fleuve noir, et maisons closes,
> Et quartiers sinistres comme des Morgues . . .

The figure of the poet, referred to in a familiar stance as 'l'Attardé' (one is reminded of the narrator of 'Soirs de fête', who arrives when the festivities are drawing to a close), is a physical component of the scene of winter desolation. Hamlet's 'Get thee to a nunnery' speech to Ophelia, which is the poem's epigraph, has already given the figure an identity, and the poem a theme. It is 'la souillure des innocentes', an obsessive theme for the idealistic Laforgue, who frequented brothels but who told his sister when announcing his engagement to Leah: 'Je ne

[32] The convent building in Tarbes that stands in this precise location had become a military barracks at the time of the Revolution.

l'ai pas encore embrassée.'[33] Throughout the poem disparate line-lengths convey the violence of the emotional swings, ranging from suspicion and jealousy to fear and concern. The free verse allows their bare and separate presentation. Successive short lines are nothing more than passionate exclamations. Long lines are followed by very brief ones, sometimes for the expected purpose of bathos, as in the following examples:

> Tout ce qui n'est pas cette vie de Vestale
> Provinciale,
>
> Mais toi, il ne faut pas que tu t'abandonnes,
> A ces vilains jeux!...

but more often for subtler effects of anxiety or tentative anticipation, as in the rewriting of Baudelaire's famous lines:

> O Nature, donne-moi la force et le courage
> De me croire en âge,

or the following equivocal understatement:

> Dans la prière, le ménage, les travaux de couture;
> Et que cela suffise...

In the third stanza of XII contrasting lengths jostle with one another with a result that is both humorous and poignant:

> Oh! si elle est dehors par ce vilain temps,
> De quelles histoires trop humaines rentre-t-elle?
> Et si elle est dedans,
> A ne pas pouvoir dormir par ce grand vent,
> Pense-t-elle au Bonheur,
> Au bonheur à tout prix
> Disant: tout plutôt que mon cœur reste ainsi incompris?

Each line-length has a psychological explanation and justification from the brief and hesitant 'Pense-t-elle au Bonheur', to the over-extended final explosion of feeling. The natural rhythms of the consciousness surface in the binary alternative conjectures and in the development of the concept of happiness from an abstract ideal to a present and urgent need. But the antithesis of 'dehors' and 'dedans', and the anadiplosis on 'bonheur' are too obvious as stylistic devices not to provoke a smile.

[33] *OC*, v. 151. The quotation continues: '—hier, j'étais assis près d'elle en voiture, dans la soirée, et en la regardant l'idée m'est venue que je pourrai caresser ses cheveux,—et j'en ai eu le vertige, et je n'en suis pas encore là—loin de là.'

Self-conscious artistry parallels psychological mockery—at the figures of a passive but jealous lover and of a childish, petulant girl. Rhyme is insistent but irregular and unsettling. 'Incompris' and 'prix' are brought together in an uneasy conjunction. The emphatic ternary rhyme, 'Temps', 'dedans', 'vent' points to an absent rhyming partner, the pervasive 'couvent' of the poem. The language is colloquial and bare, matching the asking of fundamental but unanswerable questions. Eliot considered the aim of French Symbolist poetry—as of the Metaphysical English poetry of the seventeenth century—to be 'to find the verbal equivalent for states of mind and feeling'.[34] To achieve this, Laforgue, again in Eliot's words, 'stretches, contracts and distorts the traditional French metre'.[35] This is necessary because, for him, poetry can no longer be a stylization of life; it presents the dialogue between the stylized and the random that is the subject's experience of reality: 'l'instant pris à la gorge',[36] in Mallarmé's words, or the 'stream of thought, of consciousness or of subjective life',[37] in William James's. Stanzas vary in length and character because each represents an element in this 'stream of thought'. 'Le vers libre', wrote Kahn many years later, 'est essentiellement mobile et ne doit point codifier de strophe. C'est *l'accent d'impulsion* et son appropriation à l'importance, à la durée du sentiment évoquée, ou à la sensation à traduire qui en est la déterminante.'[38] Stanza divisions echo these internal divisions, the distance between stanzas being experienced by the reader both as a disruption and as a gap that he must attempt to bridge.

William Empson, when describing the task of the reader of poetry, took two lines of Chinese poetry:

> Swiftly the years, beyond recall.
> Solemn the stillness of this spring morning.

and, stressing the reader's positive role, explained his activity thus: 'The reason why these facts should have been selected for a poem is left for him [the reader] to invent; he will invent a variety of reasons and order them in his own mind. This, I think, is the essential fact about the poetical use of language.'[39] Conversely, Jean Cohen used a fragment of

[34] T. S. Eliot, *Selected Essays* (London, 1932), 289.

[35] T. S. Eliot, Introduction to *Selected Poems of Ezra Pound* (London, 1928) p. viii.

[36] Grojnowski, *Laforgue et l'originalité*, 175. The statement was made of Dujardin's *Les Lauriers sont coupés*.

[37] W. James, *Principles of Psychology* (New York, 1902), 229.

[38] G. Kahn, 'Préface sur le vers libre', in *Premiers Poèmes* (Paris, 1897), 33.

[39] W. Empson, *Seven Types of Ambiguity* (London, 1984), 25.

prose—a newspaper item—which he set out as poetry, to illustrate how poetry demanded and inspired a different reading from that of prose. He found that set out as verse, 'les mots s'animent, le courant passe'.[40]

Hier sur la Nationale Sept
Une automobile
Roulant à cent à l'heure s'est jetée
Sur un platane
Ses quatre occupants ont été
Tués.

Both critics emphasize the author's reliance on the imagination of the reader to give substance to images and to forge the links between them.

Daniel Grojnowski employs Cohen's process in reverse. By writing the final lines of 'Pétition V' as prose,[41] he illustrates the gap between Laforgue's free verse and both prose and prose poetry:

> Oh, simplement d'infinis échanges à la fin de journées à quatre bras moissonnées, quand les tambours, quand les trompettes, ils s'en vont sonnant la retraite, et qu'on prend le frais sur le pas des portes, en vidant les pots de grès à la santé des années mortes qui n'ont pas laissé de regrets, au su de tout le canton que depuis toujours nous habitons, ton ton, ton taine, ton ton.

It has to be said that the *Derniers Vers* would not always produce such a garbled piece of prose as this, but, nevertheless, the example is an effective illustration of the value of the free-verse setting, which assists the reader and renders the text accessible. It emphasizes the fact that Laforgue's free verse is composed around the phrasing of the line, both psychologically and rhythmically. Read as free verse, we draw from this conclusion the full sense of the friendship and harmony to which the poet aspires. The phrases become invitations to imaginative conjecture, right down to the sounds of the horn, which are no longer merely cacophanous units.

Since the line as 'unit of sense' is important to Laforgue, he retains the devices that stress line divisions. Capitalization at the start of the line is one. It can also contribute to effects of bathos by the importance that it appears to attach to a trivial phrase or word: 'Vestale | Provinciale', for example. Another device is rhyme, and we have seen Laforgue use this emphatically but irregularly, exploiting to the full the suspense between the anticipated and the unpredictable that is integral to rhyme, and

[40] J. Cohen, *Structure du langage poétique* (Paris, 1966), 76.
[41] Grojnowski, *Laforgue et l'originalité*, 174.

which is intensified by the uncertain patterns of free verse. Internal rhyme and intensive liminal repetition are also used. The tenth stanza of XII will serve as an example here:

Oh! ce ne fut pas et ce ne peut être,
Oh! tu n'es pas comme les autres,
Crispées aux rideaux de leur fenêtre
Devant le soleil couchant qui dans son sang se vautre!
Oh! tu n'as pas l'âge,
Oh, dis, tu n'auras jamais l'âge,
Oh, tu me promets de rester sage comme une image? . . .

Liminal phonetic repetition is often, as here, based on a repeated exclamatory epithet. Anticipatory and apprehensive, it brings the reader back to his starting-point at each new line. It is nothing more than the registering of emotion. Clive Scott points out that, in another writer, phatic expressions such as 'Oh!' might well have been marginal 'chevilles'; in these texts they are 'a center of Laforgue's existential drama . . . the *cheville* is where the Laforguian self is locked'.[42] Paradoxically, these dramatic exclamations emphasize stasis and impasse. Like his rhetorical imperatives, they do not inaugurate change. But their immobility is balanced by the mobility of the consciousness, and we are particularly aware of this in a poem that begins with an engulfing, negative despair, and ends on an expression of tentative hope. Motifs are developed and modulated, in line with Laforgue's Wagnerian sympathies. The opening stanzas accustom the reader to the process by a witty transposition of elements: the heart and flesh of the narrator become, metaphorically, those of the girl:

. . . 'Oh! arrose, arrose
Mon cœur si brûlant, ma chair si intéressante!'

Oh, elle, mon cœur et ma chair, que fait-elle? . . .

Heart, body, and soul, the trio that we saw in 'Solo de lune', are again interwoven in a complex manner: we move from the heart, mentioned five times in the first part of the poem, to the soul—of the poet, of the nuns, or of the convent girls—and from the soul to the eyes, the traditional windows of the soul. And yet, paradoxically, the direction of the poem moves from the desire for purity towards an acceptance of life—from the soul to the heart. The girls' modest lowering of their eyes ('les yeux baissés'), expanded luxuriously in 'tes beaux yeux irréconciliable-

[42] Scott, *Vers libre*, 227.

ment baissés', indicates a rejection of life that the poet will not ultimately even desire. The poet's contrary plea to look life in the face echoes the image in reverse: 'O Nature, relève-moi le front!' A parallel opposition is met in the phrases that describe maturity. In this stanza, 'Oh! tu n'as pas l'âge' is addressed to the girl and said with apprehension. In the final stanza it will be addressed to the self and, after the alternative of deathlike isolation has been explored, it will be desired: '. . . donne-moi la force et le courage | De me croire en âge'. The repeated word is emphasized by its placing in a rhyming position (repetition is, of course, unacceptable in traditional versification): 'Oh! tu n'as pas l'âge, | Oh, dis, tu n'auras jamais l'âge'. The same technique is used in the final stanza, with the repetition of 'mourrons', again in a rhyming position. In this Eliotesque transition, we are first given the sudden, alarming apprehension: 'Oh, que nous mourrons!', isolated to form a stanza, as we have seen. Six lines later, in the poem's conclusion, the fact can be viewed with a degree of urbanity as a commonplace and as a basis for action:

> O Nature, relève-moi le front!
> Puisque, tôt ou tard, nous mourrons. . .

The mobility with which motifs are endowed is at the heart of Laforgue's method of composition. Even the central image of the winter night undergoes subtle permutations. Presented dramatically in the first two lines of the poem:

> Noire bise, averse glapissante,
> Et fleuve noir, et maisons closes,

it recurs, clipped to one line, its elements transposed, and figuring merely as a familiar background feature:

> Nuit noire, maisons closes, grand vent . . .

Later we will be reminded in more lyrical lines of its continuing and inescapable presence:

> La nuit est à jamais noire,
> Le vent est grandement triste,

and later still the narrator remembers it with sudden alarm: 'que la nuit est noire!' By this stage the 'beaux yeux' of the girl have undergone a modulation; they are now capable of deception: 'Eh bien, pour aimer ce qu'il y a d'histoires | Derrière ces beaux yeux d'orpheline héroïne.'

'Histoires' are sexual adventures ('De quelles histoires trop humaines rentre-t-elle?'), but they are not only this:

> Tout dit la vieille histoire
> Qu'il faut être deux au coin du feu . . .

The word can represent the respectable, if it still denotes the mundane, as opposed to the exceptional 'héroïque destinée'. What kind of 'histoire' and what kind of 'destinée' are we to link with the 'orpheline héroïne' of the poem, who is both Ophelia and an ordinary 'jeune fille'? The poem constantly poses such questions by means of verbal echoes that turn out to be mutations. It is clear that, despite the 'story-line' of the poem, we inhabit a shifting world of the imagination. Laforgue's 'il faut tenter de vivre'[43] may provide a concluding direction, but the poem's richness lies in the range of interrelating insights that are spread out before the reader, binary in their patterning—from the 'saintes vieilles filles' to the 'femme', the bleak night to the sexual orgy of the sunset, from 'âme' to 'créature'—but ultimately unresolvable. 'Tous les types de Rembrandt ont un air ambigu',[44] Laforgue had said approvingly, and, writing of the capturing of emotion in the visual arts: 'Dans l'émotion artistique il ne faut jamais s'en référer à l'agréable, au pénible, etc.—La lutte, l'hésitation, les conflits, la déception, la soif, tout ce qui constitue la vie doit constituer la vibration esthétique.'[45] The liberation of traditional metre gave Laforgue the freedom to cultivate these 'aesthetic vibrations' with the greatest possible accuracy and flexibility. He admired the syntactical freedoms that Mallarmé allowed himself for their ability to render states of mind: 'quant à ces absences de syntaxe, ou plutôt de fil de phrase (principe Mallarmé). Je les adore parce qu'elles sont souvent adéquates à certains états et uniquement indispensables . . .'[46] His own metrical innovations had become indispensable for the same reason.

Laforgue's much-desired innovation—there can be few writers for whom novelty was so much a self-confessed and reiterated goal—is reached by liberation and transgression. Among his most frequent transgressions is the use of the line that extends beyond the alexandrine, bearing a close relation to prose, and often used for deliberately unpoetic effects—the accumulation of well-known elements, for example. The Verlainian imparisyllabic line, which destroys binary regularity, is a technique that Laforgue uses more frequently as the *Derniers Vers*

[43] P. Valéry, 'Le Cimetière marin', *Charmes*.
[44] *MP* 170.
[45] Ibid. 176–7.
[46] *LA* 68.

progress. More of a transgression at this date is the line that is metrically ambiguous. Apollinaire begins 'Zone' with a one-line stanza, 'A la fin tu es las de ce monde ancien,' which reminds the reader metrically of the 'old' world even as it renounces it, since it can be read as a traditional alexandrine, counting 'ancien' as three syllables, or as a 'modern' imparisyllabic line, counting 'ancien' as two. Laforgue had already done something very similar in the first line of the *Derniers Vers*: 'Blocus sentimental! Messageries du Levant! . . .', where the line becomes an alexandrine if 'Messageries' is counted as three instead of four syllables. Throughout the *Derniers Vers* there are lines whose status can be changed by the use of apocope or syncope, synaeresis or diaeresis, and whose rhythmical identity depends on syllabic uncertainty. The following occur in the final twelve lines of XII:

> Tes beaux yeux irréconciliablement baissés.
>
> Que toutes sont créature, et que tout est routine!
>
> Derrière ces beaux yeux d'orpheline héroïne,
>
> O Nature, donne-moi la force et le courage . . .

Their syllabic uncertainty recalls the alexandrines that they approach, as the emotions they describe aspire to a state of harmony that is unattainable.

The rhythmical pulse of Laforgue's free verse is variable and frequently disorientating. It contains the deliberately unharmonious, as it must do, since he was, as he saw it, creating a language *for* his experience, a language that was receptive to the discontinuities and contradictions that he found in life. Cliché, parody, colloquialism, improvisation, Eliot's 'visions and revisions', must necessarily all play their part. The lyrical impulses emerge in contrast to these impediments and also by means of them. 'We are a long way indeed from a verse that has any pretensions to harmony or an aesthetic status',[47] Scott writes. The verse nevertheless contains traditional binary and ternary rhythms in single and multiple groupings, and, necessarily, also the subversion of these patterns, just as syllabic expectations are frustrated. The result might be thought to be a verse that 'reacts' too much, that is too intently turned in on the minutest sensations of its private world, as late Symbolism was accused of being. But Laforgue was too concerned with the broadest issues for this to be the case. He conceived of rhythm

[47] Scott, *Vers libre*, 239.

also in broad terms and across the poem as a whole. This is surely what he had in mind when, in 'Simple Agonie', he expressed his desire for his 'melody' to rise and portray 'la chose qu'est la chose'. Earlier he had written of his artistic ambitions: 'J'ai voulu faire de la symphonie et de la mélodie',[48] and he had admired Verlaine's *Sagesse* because it was 'de la vraie poésie, des vagissements, des balbutiements dans une langue inconsciente ayant tout juste le souci de rimer'.[49] The rhythms he sought stemmed from the unconscious; they were the essential rhythms of life, which he set in apposition to one another in the whole span of a poem, as we have seen in XII, creating 'une liaison dynamique des pulsions',[50] and thus a mobile identity from the fragmented raw materials of experience.

48 *OC*, v. 137. 49 *LA* 48.
50 J. Kristeva, *La Révolution du langage poétique* (Paris, 1974), 258.

Conclusion

WE have traced a journey from innocent or naïve poetry to a final, carefully calculated, and superficially misleading appearance of naïvety. In mid-course Laforgue spoke of 'possessing' his language ('Aujourd'hui que je . . . possède ma langue d'un façon plus minutieuse, plus clownesque . . .'[1]), and the *Complaintes* and the *Lunes* were the products of this verbal control. Irony characterized it, and allowed him to retain his artistic inscrutability and his independence of stance. As his subject-matter grew more varied and shifting, and he simultaneously tried to grapple more directly with the world of real events, he moved towards experiments in *vers libéré*, which led naturally to the invention of *vers libre*. He used it to broaden his emotional range to include a sustained intensity of feeling that had not been attempted since the early poems of the *Sanglot* period. In free verse he found the liberation that made it possible to create intricate symphonic structures, combining his sensitive aspiration to a range of ideals with his awareness of the fragmentary and the discordant. Originality had been the aim, and literary innovation was the result.

The desire for originality derived in part from an acute recognition of the multiple diverging angles of vision of the viewer, and the uniqueness of each viewer: 'Mon clavier est perpétuellement changeant et il n'y en a pas un autre identique au mien. Tous les claviers sont légitimes.'[2] Laforgue had found an illustration of this perception in the visual arts, and at one time had aimed to write a work of art criticism that would be devoted to each artist who had 'created a world', 'une série d'études où, par une accumulation de mots (*sens* et *sonorités*) choisis, de faits, de sentiments dans la gamme d'un peintre, je *donnerai* la *sensation* du monde créé par ce peintre'.[3] This personal world was not created, however, by retreating from the 'real'. Laforgue urged Kahn to cling to the 'vrai' and the 'vécu', and praised him when he did so.[4] This

[1] *AH* 821. [2] *MP* 141. [3] *AH* 687 (Laforgue's emphases).

[4] 'La définition des cheveux roux est vraie, vécue et fait bien. . . . Ce phénomène du rêve se fiançant au passé est bien vrai et vaut une romance encore que j'attends de toi.' (*LA* 68–9.) This insistence on the importance of the 'vrai' and the 'vécu' does not make Laforgue 'basically intent on mimesis' (E. A. Howe, *Stages of Self* (Athens, Oh., 1990), 117). Since Laforgue was perfectly aware that 'l'œuvre ne sera jamais l'équivalent de la réalité fugitive' (*MP* 140), 'translation' rather than 'imitation' was his aim, as it was Baudelaire's. He did, however, try to avoid what Larkin termed 'a false relation between art and life' (B. Everett, *Poets in their Time* (Oxford, 1991), 243).

was because, for Laforgue, resonance in art depended less on the pure beauty of the object perceived than on 'context', that is, on the network of complex elements with which it was linked in the perceiver's mind. Art was thus, as Baudelaire had claimed, inescapably 'modern'; in Laforgue's words, 'telle grisette de Paris, telle jeune fille de salon . . . nous fera seule sangloter, nous remuera jusqu'aux tréfonds de nos entrailles, parce qu'elles sont les sœurs immédiates de notre éphémère'.[5]

Octavio Paz stressed the importance that he attached to the modern in the lecture that he gave on receiving the Nobel Prize for poetry in 1990. 'Modernity', he said, 'has been a universal passion. Since 1850 she has been our goddess and our demoness . . . from the Symbolist period, all modern poets have chased after that magnetic and elusive figure that fascinates them. Baudelaire was the first. He was also the first to touch her and discover that she is nothing but time that crumbles in one's hands.'[6] The attempt to capture the present, with its rich inner content of memory, history, and art, to see it as a meeting-point of a fluid sensibility and a baffling and equally fluid external world, is central to Laforgue's work, and forms the substance of the advances in the interior monologue that he undertook.

Time has changed our view of Laforgue. François Ruchon's seminal early study (1924)[7] praises the *Sanglot* poems too much and pays too little attention to the originality of the *Derniers Vers* to be acceptable to a modern reader. The view held by Martin Turnell (1936), that Laforgue was technically important but emotionally immature, has also dated: 'The disparity between his technical maturity and his emotional immaturity tells heavily against him,'[8] he wrote. This view is still held by Michael Collie in his studies of the 1960s and the 1970s: 'Whatever reservations it might be necessary to have about Laforgue himself, there is no doubt that many of these poems [*Derniers Vers*] are highly accomplished works' (1963);[9] 'Had he lived longer . . . he would no doubt have arrived at a more sophisticated view of the function of art' (1977).[10]

5 *MP* 161.

6 Octavio Paz, 'In Search of the Present' (translated by Anthony Stanton), *The Times Literary Supplement*, 21–7 Dec. 1990: 1384.

7 F. Ruchon, *Jules Laforgue, sa vie, son œuvre* (Geneva, 1924). The three outstanding general French studies on Laforgue since Ruchon's are those by M.-J. Durry (1952), P. Reboul (1960), and D. Grojnowski (1988).

8 M. Turnell, 'The Poetry of Jules Laforgue', *Scrutiny*, 5 (Sept. 1936), 149.

9 M. Collie, *Laforgue* (Edinburgh and London, 1963), 94.

10 M. Collie, *Jules Laforgue* (London, 1977), 108. Collie provides valuable analyses of *Derniers Vers* I (1963) and VII (1977). Developing this approach, J. A. Hiddleston analyses each of the *Derniers Vers*, giving a vivid picture of Laforgue's 'univers imaginaire' (J. A. Hiddleston *Essai sur Laforgue et les 'Derniers Vers'* (1980)).

Such a separation of 'technique' from 'substance' may well seem particularly inappropriate when applied to a poet for whom 'ce que le poème signifie coïncide exactement avec ce qu'il est',[11] and who, as a result, created a style, in Peter Brooks's words, 'truly intimate, even in its most *cocasse* moments, with its subject matter'.[12] But it is also inappropriate, in the case of Laforgue, to consider him in terms of maturity or immaturity. The point is not that he died at 27, and was therefore only as mature as one can be in one's twenties, but that, like Keats, it was with youthful brilliance that he gave poetic form to his sensitive and often suffering response to the world.

He was too young to be one of Verlaine's *poètes maudits*, but he belonged in spirit to that iconoclastic group that pursued the 'religion' of art and, following Baudelaire, placed 'la véritable poésie'[13] beyond the realms of good and evil. Laforgue did not express himself in the dramatic terms of the *poètes maudits*, except mockingly, but the concept of maturity is largely irrelevant to him. Death, art, and love were the themes to which, like Baudelaire, he related. He did not find what Jacques Rivière termed 'le malentendu entre les sexes' 'connu et peu intéressant'.[14] But then, as Stendhal said, 'L'amour est la seule passion qui se paie d'une monnaie qu'elle fabrique elle-même.'[15]

If the standard of maturity is unsuited to Laforgue's generation of artists, it is also one by which, since his day, artists have been less and less judged. To sense its inappropriateness, one has only to think of the great figures of the twentieth century—Proust, Joyce, or Beckett, for example. Laforgue's refusal to 'conclude', so unusual even amongst avant-garde writers of the 1880s, is now accepted and championed by Absurdist theatre and the *nouveau roman*. His rigorous fidelity to the ambiguity of his personal impressions, however inadequate or trivial, is what now supports his claim to have remained modern. Labels once used—descriptively or pejoratively—such as decadent or Nihilist, seem less and less able to characterize his complexity or the particular integrity with which he approached his artistic task.

Surprisingly perhaps, Arthur Symons, the critic who, writing in

[11] M. Blanchot, writing of Mallarmé, *Faux Pas* (Paris, 1943), 128.

[12] P. Brooks, in W. Ramsey (ed.), *Jules Laforgue: Essays on a Poet's Life and Works* (Carbondale and Edwardsville, Ill., 1969), 109.

[13] 'Je sais que dans les régions éthérées de la véritable Poésie, le Mal n'est pas, non plus que le Bien.' Projected *dédicace* of *Les Fleurs du Mal*.

[14] J. Rivière and Alain-Fournier, *Correspondance*, i (Paris, 1926), 232.

[15] Stendhal, *De l'amour* (Paris, 1959), 295.

1899,[16] first introduced Laforgue's poetry to Eliot and to the Anglo-Saxon world at large, grasped very well the real seriousness of Laforgue's modernity, and the fact that it related not merely to technique, but to 'vision'. 'The old cadences,' he wrote, 'the old eloquence, the ingenuous seriousness of poetry, are all banished, on a theory as self-denying as that which permitted Degas to dispense with recognizable beauty in his figures. Here, if ever, is modern verse, verse which dispenses with so many of the privileges of poetry, for an ideal quite of its own.'[17] He saw the necessity and value of the direction that Laforgue had taken.

> He is terribly conscious of daily life, cannot omit, mentally, a single hour of the day; and his flight to the moon is in sheer desperation. He sees what he calls *L'Inconscient* in every gesture, but he cannot see it without these gestures. And he sees, not only as an imposition, but a conquest, the possibilities for art which come from the sickly modern being, with his clothes, his nerves: the mere fact that he flowers from the soil of his epoch.
>
> It is an art of the nerves, this art of Laforgue, and it is what all art would tend towards if we followed our nerves on all their journeys. There is in it all the restlessness of modern life.[18]

The concept of 'immaturity' has no place in any of this. Symons valued aspects of Laforgue that were more serious: 'there is a great pity at the root of this art of Laforgue'.[19]

How, then, can Laforgue be situated? His role in the liberation of metre, in the development of the dramatized interior monologue, and in the renewal, by their application to a modern situation, of a whole range of ironic attitudes would seem to entitle him to a clear place in the history of Symbolism. Despite these contributions, the 'heroic' view of Symbolism became one that, neglecting even Verlaine to some extent, moved from Baudelaire to Rimbaud, Mallarmé, and Lautréamont (newly discovered in the twentieth century), bold spirits who modernized poetry either by their intransigence or by the purity of their dedication to their poetic enterprise. Rimbaud, the 'voyant', was a precursor of the Surrealists; Mallarmé, the intellectual, of Valéry. It was perhaps natural that a highly acclaimed poet, who had been considered to be 'le symbole du symbolisme',[20] but who was self-deprecating and unremittingly 'quotational', should be set on one side. Now that the claims of

[16] A. Symons, *The Symbolist Movement in Literature* (New York, 1958).

[17] Ibid. 57.

[18] Ibid. 59–60.

[19] Ibid. 60.

[20] '. . . tout ce que réalisa le symbolisme, il le contient et l'exprime avec une plénitude telle (celle du génie) qu'il est vraiment le symbole du symbolisme' (E. Fagus, *La Revue blanche*, 1903).

Surrealists can themselves be seen in perspective, and the subtle and valuable role of a cultivated intertextuality is better understood, we can perhaps think again.

A contributory factor in the relative neglect of Laforgue is the absence of a theoretical position to represent him satisfactorily. Despite his constant immersion in philosophy and art criticism, and despite the vividness and acuteness of his fragmentary notes on these subjects ('Comment être aussi intelligent que Laforgue?', M.-J. Durry asked[21]), they add up even less than does Baudelaire's 'Mon cœur mis à nu' to a stance that can stand up to the theorist.[22] This is hardly surprising, but it stands in his way. It is futile, as A. G. Lehmann discovers in his examination of Symbolist theory,[23] to single out two references to 'rêve' from a number of possible candidates in Laforgue's jottings in an attempt to understand what Laforgue meant by the term that was so central to French Symbolism. Lehmann's first quotation is a whimsical observation to a female acquaintance whom Laforgue is trying to impress,[24] and his second represents the climax of an emotional celebration of the unconscious in art;[25] neither is as helpful as certain more explanatory statements that we have met in letters to friends. One has a much better sense of Laforgue's aesthetic aims by observing the poetry in action. We do not even have to turn to one of the late poems on which this study has concentrated to shed some light on the subject. 'Complainte sur certains ennuis', like most of the *Complaintes*, is a poem that still seems modern:

> Un couchant des Cosmogonies!
> Ah! que la Vie est quotidienne...
> Et, du plus vrai qu'on se souvienne,
> Comme on fut piètre et sans génie...
>
> On voudrait s'avouer des choses,
> Dont on s'étonnerait en route,
> Qui feraient une fois pour toutes!
> Qu'on s'entendrait à travers poses.

[21] M.-J. Durry, *Jules Laforgue* (Paris, 1952), 83.

[22] See M. Dottin, 'Laforgue salonnier', *Europe*, May 1985; and M. Hannoosh, 'The Poet as Art Critic: Laforgue's Aesthetic Theory', *Modern Languages Review*, 79 (July 1984).

[23] A. G. Lehmann, *The Symbolist Aesthetic in France* (Oxford, 1950), 48–9, 115–20.

[24] 'Enfin, peut-être tout n'est-il qu'un rêve, seulement celui qui nous rêve ferait bien de hâter le cuvage de son opium.' (Ibid. 99; *AH* 763; Letter to Mme Mültzer, Mar. 1882.)

[25] '... ne sachons que nous enivrer des paradis sans fond de nos sens et fleurir sincèrement nos rêves sur l'heure qui est à nous' (Lehmann, *Symbolist Aesthetic*, 100; *MP* 207–8 (from 'L'Art moderne en Allemagne')).

On voudrait saigner le Silence,
Secouer l'exil des causeries;
Et non! ces dames sont aigries
Par des questions de préséance.

Elles boudent là, l'air capable.
Et, sous le ciel, plus d'un s'explique,
Par quels gâchis suresthétiques
Ces êtres-là sont adorables.

Justement, une nous appelle,
Pour l'aider à chercher sa bague,
Perdue (où dans ce terrain vague?)
Un souvenir D'AMOUR, dit-elle!

Ces êtres-là sont adorables!

Its title already establishes that the tone of the poem will be low-key. We are not being offered Romantic ennui or Baudelairean spleen, but 'certains ennuis'—a number of minor irritations and inconveniences. They turn out, ironically, to be the eternal human problems, those that Baudelaire treated: the inability to create; the still-born aspiration towards total communication in love; the human condition, with its awareness of futility, banality, and isolation; but they sadly lack the grandeur they had once possessed. The poet-narrator is self-avowedly 'piètre et sans génie'. And yet the poem is delightful and, in its final line, even debonair. All of this is typically Laforguian. Ezra Pound praised Laforgue's combination of power and nonchalance. If the reason 'why such power should coincide with so great a nonchalance of manner'[26] remains mysterious, we are clear that this is the combination that we are being offered and that it is a combination that is particularly appropriate to the modern world. This poem would seem to have little claim to power were it not for its constant ability to surprise, to shift the reader's perception of things, to make him see the world through the poet's eyes. Laforgue's early references to bleeding hearts, silence, or exile had lacked the force that the taut metaphors of the following two lines possess:

On voudrait saigner le Silence,
Secouer l'exil des causeries . . .

Laforgue's style is, like Philip Larkin's, 'wilfully modest' and 'peculiarly potent',[27] a combination that one connects with post-Symbolism more

[26] E. Pound, *Literary Essays* (London, 1954), 283.
[27] B. Everett, *Poets in their Time*, 243.

than with Symbolism, for, as Barbara Everett points out, Symbolism is arrogant, whereas the humbler literature that follows it is not.

Laforgue becomes a master of elliptical 'collage' techniques. In the phrase 'du plus vrai qu'on se souvienne', for example, 'vrai' replaces the expected 'loin', and the alien epithet first startles and then pleases by its appropriateness. Because we *are* concerned, not with distance or nostalgia, but with accuracy of vision, and it is amusing to have the fact picked out with such economy. This substitution prepares us for the surprise repetition, in contrasting register, of the final line, which reverses our view of the lonely narrator. He is, after all, capable of responding to the charm of the society women, idle though their chatter is, and vain as they clearly are. Emotion and intellect, as Eliot pointed out, do not sit comfortably together. And the paradox has to be seen as amusing, that it is the frivolous women who supply the poem with its image of love and of the poetic—about both of which they know in reality so little—while the poet remains, in Keats's words, 'the most unpoetical of all God's Creatures',[28] as he pursues his voyeuristic path, an outsider to the romance of life to which he aspires with such sensitive and intelligent dedication, and the ambiguity of which he captures for the reader by his disconcertingly rigorous refusal of false sentiment.

We have to move into the twentieth century for the best parallels to such delicate handling of the vagaries of human emotions: the French ironic tradition, which includes Apollinaire, Léon-Paul Fargue, and Supervielle, but perhaps even more the Anglo-Saxon tradition.[29] Eliot comes to mind naturally, since the *complainte* recalls elements of poems such as 'Portrait of a Lady' (and contains the title of *The Waste Land*), but also poets such as Pound, Wallace Stevens, Hart Crane, Auden, and Larkin.

Laforgue created a way of seeing the world, which, once experienced, could not easily be shaken off. Francis de Miomandre put the point wittily in verse:

[28] J. Keats, *Letters*, ed. M. Buxton Forman (Oxford, 1935), 228.

[29] This may be an Anglo-Saxon point of view. M.-J. Durry sees his influence in all of the following (some in their beginnings or burlesque moods only): Gide, Toulet, Derême, Pellerin, Carco, Jean de Tinan, Marcel Schwob, Marcel Achard, Prévert, Anouilh, Giraudoux, Claudel (in *Protée* and *L'Ours et la lune*), Jarry, Apollinaire, Léon-Paul Fargue, Michaux, Queneau (Durry, *Jules Laforgue*, 13).

> C'était un jeune homme comme tous les autres,
> Mais il avait beaucoup lu Laforgue
> Et il en reste toujours quelque chose.[30]

Pound put it not from the reader's angle but from that of the writer, in his insistence that, after Laforgue, certain ways of describing experience were no longer possible. Acutely aware of the inadequacy of received ideas and of received forms, Laforgue described the present as an amalgam of past and future projections. He located experience in the gaps between accepted literary modes, and, in particular, between the comic and the tragic muses. He conceived of selfhood as inescapably multiple, and offered his reader images of a gallery of partial selves. As a simple consequence of these things, he was the first poet to capture the fleeting inconsistencies of human emotion in appropriately free verse. His experience was not embraced in, or explained by, any comprehensive view. If he had lived another thirty years, it is, of course, possible that he might have changed his attitude to life and art, becoming more like the moralists who preceded him, and less like the giant decadents of the twentieth century who succeeded him. He did not. His achievement was the creation of a poetic method that was entirely appropriate to the inconsequentiality of experience, which the artist has, over the last hundred years, particularly sought to convey. In that sense, he was, as far as poetry was concerned, quite simply the first of the moderns.

Just as Gertrude Stein grew to look more like the portrait that Picasso painted of her, so Laforgue's image of a world defined by the heterogeneous and the inconstant has become a more easily recognizable one. It is the clarion-calls of trail-blazers into the modern—Rimbaud, or the Surrealists—which may now seem dated in their extremism or optimism: 'La première étude de l'homme qui veut être poète est sa propre *connaissance, entière*';[31] 'le surréalisme poétique . . . s'est appliqué . . . à rétablir dans sa *vérité absolue* le dialogue'[32] (my emphases). The 'decadent' outsider's portrayal of the fractured consciousness is, on the other hand, still fresh—

30 F. de Miomandre, 'Un ami de Laforgue', *L'Occident*, Dec. 1903.

31 A. Rimbaud, 'Lettre du voyant' to Paul Demeny, May 1871, *Œuvres complètes* (Paris, 1951), 254.

32 A. Breton, 'Manifeste du surréalisme', 1924 (Paris, 1985), 47.

Quand j'organise une descente en Moi,
J'en conviens, je trouve là, attablée,
Une société un peu bien mêlée,
Et que je n'ai point vue à mes octrois.[33]

—as is his expression of the poet's perennially unsatisfied desire for meaning and synthesis through art:

Oh! que
Devinant l'instant le plus seul de la nature,
Ma mélodie, toute et unique, monte,
Dans le soir et redouble, et fasse tout ce qu'elle peut
Et dise la chose qu'est la chose,
Et retombe, et reprenne,
Et fasse de la peine,
O solo de sanglots,
Et reprenne et retombe
Selon la tâche qui lui incombe.[34]

[33] 'Ballade'. [34] 'Simple Agonie'.

Bibliography

THE WORKS OF LAFORGUE (in chronological order of publication)

Les Complaintes (Paris, 1885).

L'Imitation de Notre-Dame la lune (Paris, 1886).

Le Concile féerique (Paris, 1886).

Moralités légendaires (Paris, 1887).

Les Derniers Vers (including *Des Fleurs de bonne volonté*, *Le Concile féerique*) ed. E. Dujardin and F. Fénéon (Paris, 1890).

Poésies complètes (Paris, 1894).

Œuvres complètes, 3 vols. (Paris, 1902–3): vols. i and ii contain poems; vol. iii, *Mélanges posthumes.*

Œuvres complètes, ed. G. Jean-Aubry (Paris, 1922–30): i, ii. *Poésies*; iii. *Moralités légendaires*; iv, v. *Lettres*; vi. *En Allemagne.*

Lettres à un ami (1880–6), ed. G. Jean-Aubry (Paris, 1941).

Stéphane Vassiliew, ed. F. Ruchon (Geneva, 1946).

Les Complaintes, L'Imitation de Notre-Dame la lune, Derniers Vers, ed. Cl. Pichois (Paris, 1959).

Derniers Vers, ed. M. Collie and J. M. L'Heureux (Toronto, 1965).

Poésies complètes, ed. Pascal Pia (Paris, 1970).

Poems (selected), ed. J. A. Hiddleston (Oxford, 1975).

Mélanges posthumes, ed. P. Bonnefis (Geneva, 1979).

Moralités légendaires, ed. D. Grojnowski (Geneva, 1980).

Feuilles volantes, ed. D. Grojnowski (Paris, 1981).

Œuvres complètes, i, ed. J.-L. Debauve, D. Grojnowski, P. Pia, and P. O. Walzer (Lausanne, 1986). This edition will consist of 3 vols.: vol. i contains: *Tessa*, *Chroniques* from *La Guêpe* and *L'Enfer*, Early Poems, *Proses* from *La Vie moderne*, *Le Sanglot de la terre*, *Stéphane Vassiliew*, *Amours de la quinzième année*, *Pierrot fumiste*, *Les Complaintes*, Notes (1880), Correspondence (1878–83), *Agenda* (1883).

Textes de critique d'art, ed. M. Dottin (Lille, 1988).

Fuller bibliographies may be found in the following works: F. Ruchon, *Jules Laforgue, sa vie, son œuvre* (Geneva, 1924); W. Ramsey, *Jules Laforgue and the Ironic Inheritance* (New York, 1953); D. Grojnowski, *Jules Laforgue et l'originalité* (Neuchâtel, 1988).

WORKS CONCERNED WITH LAFORGUE

These are listed in chronological order of publication in order to indicate roughly the process of evolution of Laforgue criticism.

M. Dufour, *Etude sur l'esthétique de Jules Laforgue* (Paris, 1904).
F. Ruchon, *Jules Laforgue, sa vie, son œuvre* (Geneva, 1924).
L. Guichard, *Jules Laforgue et ses poésies* (Paris, 1950).
M.-J. Durry, *Jules Laforgue* (Paris, 1952).
W. Ramsey, *Jules Laforgue and the Ironic Inheritance* (New York, 1953).
P. Reboul, *Laforgue* (Paris, 1960).
M. Collie, *Laforgue* (Edinburgh, 1963).
W. Ramsey (ed.), *Jules Laforgue: Essays on a Poet's Life and Work* (Carbondale and Edwardsville, Ill., 1969).
J.-L. Debauve, *Laforgue en son temps* (Neuchâtel, 1972).
M. Collie, *Jules Laforgue* (London, 1977).
M. Betts, *L'Univers de Laforgue à travers les mots* (Paris, 1978).
D. Arkell, *Looking for Laforgue: An Informal Biography* (Manchester, 1979).
P. Bonnefis (ed.), *Jules Laforgue*, in *Revue des sciences humaines*, 178 (1980).
J. A. Hiddleston, *Essai sur Laforgue et les 'Derniers Vers' suivi de Laforgue et Baudelaire* (Lexington, Ky., 1980).
L. Watson, *Laforgue: Poet of his Age* (Mahwah, NJ, 1980).
Jules Laforgue (a special issue devoted to Laforgue), in *Europe, revue littéraire mensuelle* (May 1985).
D. Grojnowski, *Jules Laforgue et l'originalité* (Neuchâtel, 1988).
J. A. Hiddleston (ed.), *Laforgue aujourd'hui* (Paris, 1988).
M. Hannoosh, *Parody and Decadence: Laforgue's 'Moralités légendaires'* (Columbus, Oh., 1989).

ARTICLES ON LAFORGUE (in alphabetical order)

C. Abastado, 'La Glace sans tain', *Littérature*, 27 (Oct. 1977).
—— '*Préludes autobiographiques* de Jules Laforgue: Anamorphose d'un mythe et dérive d'une écriture', *Romantisme*, 39 (1983).
J. Bem, 'Le Blanc de la complainte', *Nineteenth-Century French Studies* (Autumn-Winter 1986–7).
G. Blin, 'A la recherche de l'infini: Laforgue et Baudelaire', *Revue hebdomadaire* (5. Nov. 1938).
R. R. Bolgar, 'The Present State of Laforgue Studies', *French Studies* (July 1950).
—— 'Laforgue's *Derniers Vers*: Some Textual and Critical Problems', *Forum for Modern Language Studies*, 3: (1967).
U. Franklin, 'Reverberations of Heine in Laforgue's Lyrics', *Comparative Literature* (Spring 1987).
D. Grojnowski, 'Poétique de Laforgue', *Critique*, 237 (Feb. 1967).
—— 'Poétique du vers libre: "Derniers Vers" de Jules Laforgue', *Revue d'histoire littéraire de la France* (Oct. 1984).
—— 'Laforgue fumiste: l'esprit de cabaret', *Romantisme*, 64 (1989).

M. Hannoosh, 'The Poet as Art Critic: Laforgue's Aesthetic Theory', *Modern Languages Review*, 79 (July 1984).

—— Metaphysicality and Belief: Eliot on Laforgue, *Comparative Literature* (Autumn 1987).

J. A. Hiddleston, 'Dans l'intertexte laforguien: Hartmann et quelques autres', *Romantisme*, 64 (1989).

E. Howe, 'Repeated Forms in Laforgue', *Nottingham French Studies*, 24 (1985).

M. Jarrety, '"Et je passe et les abandonne": Poétique vagabonde de Laforgue', *Romantisme*, 64 (1989).

Y. Leclerc '"X en soi?" Laforgue et l'identité', *Romantisme*, 64 (1989).

S. Mori, 'Poétique du silence ou le principe de la composition chez Jules Laforgue', *Études de langue et de littérature française* (Société japonaise de langue et de littérature française), 42 (1983).

GENERAL STUDIES

R. de Casteras, *Avant le chat noir les Hydropathes 1878–80* (Paris, 1945).

A. Dinar, *La Croisade symboliste* (Paris, 1943).

E. Dujardin, *Les Premiers Poètes du vers libre* (Paris, 1922).

E. J. H. Greene, *T. S. Eliot et la France* (Paris, 1951).

M. Hamburger, *The Truth of Poetry: Tensions in Modern Poetry from Baudelaire to the 1960s* (London, 1969).

J. P. Houston, *French Symbolism and the Modernist Movement: A Study of Poetic Structures* (Baton Rouge, La., 1980).

E. A. Howe, *Stages of Self: The Dramatic Monologues of Laforgue, Mallarmé and Valéry* (Athens, Oh., 1990).

G. Kahn, *Symbolistes et décadents* (Paris, 1902).

D. Kelley, *Unreal City* (Manchester, 1985).

A. G. Lehmann, *The Symbolist Aesthetic in France* (Oxford, 1950).

G. Martin (ed.), *Eliot in Perspective* (London, 1970).

P. Martino, *Parnasse et symbolisme* (Paris, 1947).

G. Michaud, *Message poétique du symbolisme* (Paris, 1947).

P. Newman-Gordon, *Corbière, Laforgue, Apollinaire ou le rire en pleurs* (Paris, 1957).

H. Peyre, *Qu'est-ce que le symbolisme?* (Paris, 1974).

M. Raymond, *De Baudelaire au surréalisme* (Paris, 1933).

J.-P. Richard, *Pages paysages: Microlectures* (Paris, 1984).

A. M. Schmidt, *La Littérature symboliste* (Paris, 1942).

C. Scott, *A Question of Syllables* (Cambridge, 1986).

—— *Vers libre* (Oxford, 1990).

J. Starobinski, *Portrait de l'artiste en saltimbanque* (Geneva, 1970).

R. Storey, *Pierrot: A Critical History of a Mask* (Princeton, NJ, 1978).

—— *Pierrots on the Stage of Desire* (Princeton, NJ, 1985).

A. Symons, *The Symbolist Movement in Literature* (London, 1899).

Index